A-Level CHEMISTRY

Materials Science

E.N. Ramsden

B.Sc., Ph.D., D.Phil.

Formerly of Wolfreton School, Hull

Stanley Thornes (Publishers) Ltd.

First published 1995 by
Stanley Thornes (Publishers) Ltd,
Ellenborough House,
Wellington Street,
CHELTENHAM,
GL50 1YD

A catalogue record for this book is available from the British Library.

ISBN 0 7487 1807 9

The front cover shows a polarised light micrograph of a thin layer of nickel oxide

Acknowledgements

I offer my sincere thanks to Professor S.B. Palmer for reading the sections on engineering properties of materials and to Dr E.S. Albone and Mr D.J. Gradwell for reading the first draft and making valuable comments.

I am grateful to the following people and firms for supplying photographs:
Aviation Picture Library: 5.4A
ICI: 4.17E, 4.19B
Lotus Cars Ltd: 5.10A
M.G. Duff Marine Ltd: 2.13A
Martyn Chillmaid: 4.1A, 4.9C
Mass Transit Railway Corporation, Hong Kong: 2.1A
NASA: 3.1A
Orange Mountain Bikes Ltd: 6.7F
Pilkington plc: 3.7B
Rolls-Royce plc: 2.6D
Science Photo Library:
1.13B (Tim Beddow); 2.4A (Labat, Jerrican); 2.9G (Manfred Kage); 4.16A (Peter Meazel); 4.17A (Alex Bartel); 5.8D (Dr Jeremy Burgess); cover

I thank the publishing team who have contributed their enthusiasm and expertise to the production of this book, especially Adrian Wheaton as Science Publisher and John Hepburn as Editor.

My family have given me their support and encouragement all through the writing of this book.

E.N. Ramsden,
Oxford, 1995

Typeset by Tech-Set, Gateshead, Tyne & Wear.
Printed and bound in Great Britain at Scotprint, Musselburgh

CONTENTS

CHAPTER 5: COMPOSITE MATERIALS

CHAPTER 6: REVIEW OF MATERIALS

PREFACE

Materials Science has been written to match the draft syllabuses for the following 1996 A-level modules:

Northern Examinations and Assessment Board
    Module Ch 7: The Chemistry of Modern Materials
University of Cambridge Local Examinations Syndicate
    Module 4823: Materials
University of London Examinations and Assessment Council
    Nuffield: Materials Science
University of Oxford Delegacy of Local Examinations
    9855/44 Module 4 Option: Polymers and Plastics
Oxford and Cambridge Schools Examination Board
    9662 Polymers Module

Before embarking on an optional topic such as *Materials Science,* students will have completed the A-level Chemistry core modules, covering atomic structure, the chemical bond and a firm foundation of physical, inorganic and organic chemistry. Should they need to revise this core material, they can consult the references to *ALC* which relate to my text, *A-Level Chemistry,* Third Edition (STP). They give the section of this text in which the relevant core material can be found. Students who are using a different A-level textbook need to consult the index of their book to find the corresponding material.

1

PROPERTIES OF MATERIALS

1.1 SOME FAMOUS FAILURES

The King's Bridge in Melbourne, Australia, opened in 1962. Fifteen months later, one of the spans collapsed as a 45 tonne vehicle was crossing. Did the engineers who built the bridge use the right material for the job? Examination showed that four girders had brittle fractures. (Brittle fracture will be discussed in §1.11.) All of these fractures had started in welds. Analysis showed that the steel used to make the failed girders was outside the specifications: it contained more carbon and manganese than the engineers had specified. This composition made it more difficult to make crack-free welds. (The importance of cracks will be covered in §1.10.) Fracture occurred on a cold night when the temperature was about 2 °C. (You will meet the effect of temperature on strength in §1.11.)

Why did these structures fail?

The *Sea Gem* was an off-shore oil drilling rig in the North Sea. It was a rectangular platform supported by twenty steel tubular legs. In 1965 the oil rig collapsed and 19 people died. This happened as the legs were being raised from the sea bed to free the pontoon so that it could be towed to another location. Investigation showed that the cause of the failure was brittle fracture of the bars which transferred the weight of the pontoon to the supporting legs. Before being in the North Sea, the rig had been used in the Middle East where temperatures are much higher. As a result of being in the North Sea, the tie bar material was brittle rather than ductile. (The transition of metals from ductile to brittle condition is covered in §1.11.) Examination of the tie bars showed narrow regions which gave a stress concentration factor of 7. (You will find out about stress concentration factor in §1.10.) This factor, together with the brittle state of the metal, was responsible for the material failing during the stresses of the operation.

... The King's Bridge in Australia ...
... the Sea Gem oil rig in the North Sea ...
... the Kohlbrandt Bridge in Germany, where signs of corrosion were spotted in time ...
... and the two Comet aircraft which crashed

The Kohlbrand Bridge is a suspension bridge, 520 m long, which links the port of Hamburg with a motorway. It was built in 1974, and by 1976 the first signs of corrosion were detected in the suspension cables. Attempts were made to protect the cables with a plastic sealant. In 1978 it was found that rain had permeated the cable strands from top to bottom and that salt applied to the road in winter had worked its way into the stays. All the cables were replaced with galvanised cables (which protect against corrosion, as explained in §2.14) to give better protection against salt.

In 1953 and 1954 crashes occurred of the relatively new aircraft, the Comet. This aircraft was one of the first to have a pressurised fuselage, and it was built from an aluminium alloy. Investigators dredged segments of a crashed Comet from the Mediterranean and fitted together the pieces. They noted a small hole near the corner of a window, spreading from which were cracks. They deduced that this fatigue damage (for an explanation of fatigue see §1.12) might be responsible for the failure. To test the theory, a cabin section was subjected to pressure tests which simulated the effects of flights. Each time an aircraft climbs, the pressure difference between the inside and the outside increases, and each time it descends the pressure difference

A study of materials science will tell you!

1

decreases. The fuselage tested had already made 1230 pressurised flights, and it failed after a further 1830 simulated flights. Failure occurred in a similar place to that in the crashed aircraft and started from a small area of fatigue damage. The investigation showed that the fundamental cause of the aircraft crash was fatigue failure, which occurred at a small hole near a window, the small hole acting as a stress-raiser. (The effect of a stress-raiser on failure is discussed in § 1.10.) A crack then started and spread slowly until it reached the critical length [see § 1.10].

1.2 CHOICE OF MATERIALS

The enormous variety of materials that you see around you includes pure metals, alloys, ceramics, plastics, wood, stone, concrete and composite materials. Some materials are widely used for a variety of purposes, but others are used for a limited number of applications. The choice of material for a particular purpose depends on:

- the conditions under which the product is used, e.g. whether it needs to be very strong, and whether it needs to be corrosion-resistant
- the requirements of the method of manufacture, e.g. whether a material can be moulded into a complicated shape or bent without breaking
- the cost.

A material is chosen for an application depending on its properties e.g. physical properties, chemical properties, subjective properties, mechanical strength, hardness and toughness

Some of the **physical properties** which are considered are density, tensile strength, toughness, hardness, electrical conductivity and thermal conductivity. These physical quantities can be measured accurately. **Chemical properties** decide whether a material can be used in a certain environment. Other **subjective properties**, such as appearance and feel, are important, e.g. in the case of a clothing material, but may not be measurable in scientific terms.

Materials which are to be used as structural engineering materials must be strong, tough and hard. **Mechanical strength** is measured (as described in § 1.5) by the **yield stress** – the stress needed to cause a permanent change of shape – or by the **fracture stress** – the stress needed to break the material. **Tensile strength** and **compressive strength** are discussed in §§ 1.4–6. **Hardness**, the resistance of a material to abrasion, is discussed in § 1.7 and **toughness**, the resistance of a material to fracture, in § 1.9.

For many applications, electrical properties [see § 1.13] and thermal properties [see § 1.14] are important in determining the choice of material for a task.

Metals, alloys, ceramics, polymers and composite materials will be considered.

The materials considered in this book are **metals and alloys** [Chapter 2], **ceramics**, e.g. porcelain and silicon carbide [Chapter 3], **polymers**, e.g. poly(ethene) and nylon [Chapter 4] and **composite materials** [Chapter 5]. Composite materials consist of a matrix of one material reinforced with fibres or rods or particles of another material. Examples are reinforced concrete and glass-fibre-reinforced polyester. The different types of materials are compared in Chapter 6.

1.3 STRESS

Mechanical properties – the way materials behave when forces are applied to them – in tension, in compression, in shear

The ways in which materials behave when forces are applied to them are described as **mechanical properties**. When an external force is applied to a material, internal forces are set up in a direction that opposes the external force. For example, when a spring is stretched by external forces, internal forces are set up in opposition to the stretch; when the spring is released these forces cause it to contract. A material which is being stretched by external forces is in **tension**. A material which is being squeezed by external forces is in **compression**. A material which is subject to forces which twist it or which make one face slide relative to another is in **shear**.

FIGURE 1.3A
(a) Tensile Forces,
(b) Compressive Forces,
(c) Shear Forces

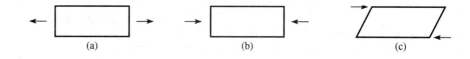

(a)   (b)   (c)

At this point some definitions will be useful.

STRESS

Stress – the force per unit cross-sectional area of material that produces deformation – includes bulk stress and tensile stress

Materials used in construction must be able to withstand the application of considerable forces while undergoing only minor deformation (change of shape). The engineer is interested in the force which is necessary to produce a definite amount of deformation, either temporary or permanent, in a material. The quantity **stress** is the force acting per unit cross-sectional area of the material, the area being measured in the direction at right angles to the line of action of the force. The proportional deformation that results is the strain (see below).

BULK STRESS

Bulk stress or **hydrostatic stress** is stress which tends to compress the whole body. It is simply the pressure on the body.

Bulk stress = Applied force/Area

TENSILE STRESS

Tensile stress is stress which tends to stretch a material in a particular direction.

Tensile stress, σ = Force in a certain direction/Original cross-sectional area

Stress has the unit of pressure, newton per square metre, $N\,m^{-2}$ or pascal, Pa. It is often expressed in megapascals, MPa, or gigapascals, GPa.

$$1\,GPa = 10^3\,MPa = 10^6\,kPa = 10^9\,Pa = 10^9\,N\,m^{-2}$$

STRAIN

Strain – the elongation that results from tensile stress
A definition of yield stress is given ..
... and fracture stress ...
... and toughness

The elongation that results from tensile stress is called **strain**.

Tensile strain = Elongation per unit length

Since tensile strain is a ratio of two lengths, it is dimensionless. It is often expressed as percentage elongation.

YIELD STRESS

The stress that produces a permanent plastic deformation in a material is called the **yield stress** [see Figure 1.5B].

FRACTURE STRESS

Non-metals tend to fracture before they change shape and for them it is more useful to quote **fracture stress** – the stress needed to break the material.

TOUGHNESS

The resistance of a material to fracture is called **toughness**.

1. Calculate the stress when a rod of material of cross-sectional area 1×10^{-6} m^2 is subjected to a tensile force of 100 N.

2. Calculate the stress when a rod of material of cross-sectional area 40 mm^2 is subjected to a tensile force of 120 N.

3. Calculate the strain when a metal wire 60 cm long is extended to 60.6 cm.

4. What is the strain when a rod of material 2.10 m long is extended by 63 mm?

1.4 STRENGTH

The tensile strength of a material is the maximum tensile stress that it can withstand without fracturing.

The **strength** of a material is its ability to resist the application of forces without breaking. The **tensile strength** of a material is defined as the maximum tensile stress which it can withstand without breaking. It has the unit of stress, N m^{-2} or Pa.

1.4.1 DEFORMATION

Deformation can be elastic or plastic or a fracture. Materials that fracture easily are described as brittle.

When a force is applied to a material, the material may be deformed. The deformation may be **elastic** or **plastic** or may be a **fracture**. **Elastic behaviour** means returning to the original size and shape when the force which is producing the deformation is removed, e.g. a spring extending when loaded and returning to its original length when the load is removed. **Plastic behaviour** means that the material retains a permanent change of shape when the force is discontinued, e.g. an electrical wire being bent into the shape needed in a circuit.

1.4.2 BRITTLE AND DUCTILE MATERIALS

Materials that can be deformed extensively before breaking are described as ductile.

A characteristic of metals is that, when sufficiently stressed, they deform plastically, while most non-metallic substances do not. A material which can withstand only a small amount of deformation before fracture is **brittle**; one which can be deformed extensively before it breaks is **ductile**. The ductility of metals, which arises from the nature of the metallic bond [see § 2.3], allows them to be shaped by processes such as rolling into sheets, hammering and drawing into wire. It also enables metals to be used in applications where brittle materials would shatter in use. The strength of the bonds between metal atoms makes metals strong and tough. Many non-metallic materials are strong in compression, e.g. ceramics, but they are brittle. Such materials can bear a load but they cannot withstand deformation without shattering.

1.5 STRESS AND STRAIN

1.5.1 THE TENSILE TEST

To investigate the manner in which a material, e.g. a metal, changes when it is stressed, the **tensile test** is carried out. The principle is to apply a force to a specimen and measure the extension that results [see Figure 1.5A]. The results of a typical set of measurements are plotted in Figure 1.5B.

FIGURE 1.5A
A Tensile Strength
Apparatus

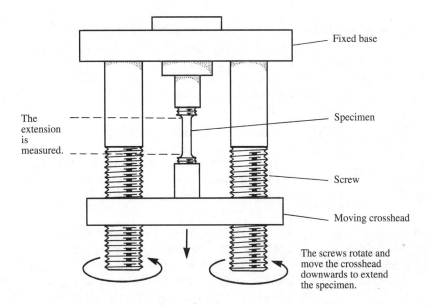

FIGURE 1.5A
A Tensile Strength
Apparatus

FIGURE 1.5B
Stress–Strain Graph for a
Specimen on Loading
(not to scale)

*The tensile test examines
the changes in a specimen
of material under tensile
stress.*

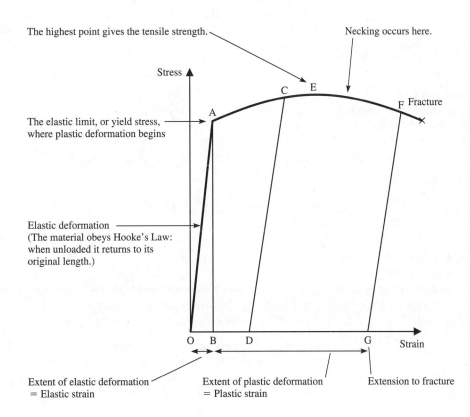

1.5.2 ELASTIC LIMIT AND YIELD STRESS

The first stage of the stress–strain curve, OA, shows elastic deformation; that is when the load is removed the specimen returns to its original length. The extent of elastic deformation is the length OB. At A, plastic deformation begins: that is, the specimen does not return to its original length when the load is removed. The stress at point A is called the **elastic limit**. In most cases, at almost the same stress the material starts to yield – to stretch without further increase in force. The stress

The strain increases as the stress increases ... through the elastic limit ... ... through the yield stress ... to the value at which fracture occurs.

at which this happens is the **yield stress**. If the specimen is stretched to C and then unloaded, unloading follows the line CD. The line CD is parallel to AO because the elastic deformation OB is reversed and the permanent extension BD is retained. Plastic behaviour continues until the load is great enough to cause **fracture** (breaking into two or more pieces).

As stress increases from A to E, the specimen is stretched and its cross-sectional area decreases. The increase of strain with stress in the plastic range between A and E is termed **work hardening**. The specimen has increased in length and strength sufficiently to carry the load. Work hardening is due to a change in the structure of the specimen caused by plastic deformation [see § 2.7].

In the plastic range of strain, a change in structure called work hardening occurs.

Between A and E, the elongation is uniform along the length of the specimen [Figure 1.5C (a)]. At E, the increase in strength due to work hardening fails to compensate for the decrease in cross-sectional area of the specimen. The deformation becomes unstable and is confined to one part of the specimen, forming a **neck** [see Figure 1.5C]. The necked region cannot transmit the load along the specimen and therefore the neck fractures soon afterwards, at point F. Point E is called the **ultimate tensile stress** or the **tensile strength**. The maximum stress which a material can withstand in compression is the **compressive strength**, which in most cases is not the same as the tensile strength.

FIGURE 1.5C Necking as a Result of Loading

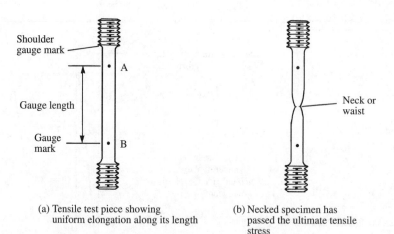

After the ultimate tensile strength has been reached, necking takes place ... ... and soon after this the specimen fractures.

(a) Tensile test piece showing uniform elongation along its length

(b) Necked specimen has passed the ultimate tensile stress

The stress–strain curve for a brittle material is shown in Figure 1.5D. The stress–strain curve for some metals which do not benefit from work hardening, e.g. zinc, tin and lead, is shown in Figure 1.5E. The stress–strain curve for mild steel, which has two yield points, is shown in Figure 1.5F

FIGURE 1.5D
The Stress–Strain Curve for a Brittle Material

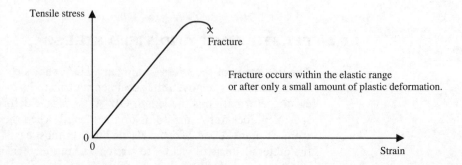

Brittle and ductile metals show stress–strain curves of different shapes.

Fracture occurs within the elastic range or after only a small amount of plastic deformation.

FIGURE 1.5E

The Stress–Strain Curve
for Metals such as Zinc,
Tin and Lead

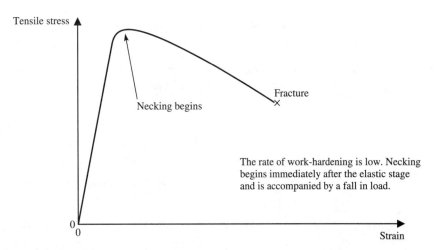

FIGURE 1.5F

The Stress–Strain Curve
for Mild Steel

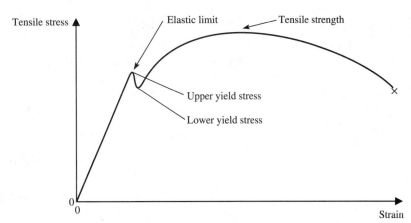

1.5.3 HOOKE'S LAW

Over the elastic part of the deformation (OA in Figure 1.5B) the specimen follows **Hooke's Law**. This states that the tension in a spring or a stretched specimen is proportional to its extension from its natural length. The tension is the force which is tending to restore the specimen to its original length. This law can be expressed by the equation:

Hooke's Law states that the tension in a spring or stretched wire is proportional to its extension from its natural length.

Restoring force = Force constant × Extension

$$F = kx$$

where x = extension, k = force constant, F = restoring force.

CHECKPOINT 1.5

1. Which of the following metals is the strongest in tension?
Metal **A** tensile strength 130 MPa
Metal **B** tensile strength 140 MPa
Metal **C** tensile strength 150 MPa

2. Calculate the tensile force that will be required to cause yielding of a bar of material with a cross-sectional area of 100 mm² and yield stress 150 MPa.

3. A steel has a yield strength of 600 MPa. What tensile force is required to make a steel rod of cross-sectional area 400 mm² yield?

4. A cast iron has a compression strength of 600 MPa and a tensile strength of 200 MPa. What does this suggest about the uses to which this material will be put?

1.6 ELASTIC MODULUS

The **elastic modulus** or **modulus of elasticity** is a property of materials. It expresses the extent to which the material changes shape when it is stressed. For a body which obeys Hooke's Law, the **elastic modulus** is given by

Elastic modulus = Stress/Strain

There are different types of modulus, depending on the type of stress. When the stress is a tensile (stretching) stress, the **tensile modulus**, known as **Young's modulus**, is given by:

Young's modulus (tensile modulus) E = Tensile stress/Tensile strain

The unit is the same as stress since strain has no unit. Values are often quoted in GPa.

One pascal = one newton per square metre; $1\,Pa = 1\,N\,m^{-2}$

One gigapascal = 10^3 megapascal, = 10^6 kilopascal = 10^9 pascal;

$1\,GPa = 10^3\,MPa = 10^6\,kPa = 10^9\,Pa$

For a body which obeys Hooke's Law, Elastic modulus = Stress/Strain The tensile modulus is known as Young's modulus. It measures the stiffness of a material.

A steel may have a modulus of 200 GPa; an aluminium alloy may have a value of 70 GPa. **Stiffness** is the ability of a material to resist elastic deformation. A stiff material has a high modulus of elasticity. From the values, you can see that steels are stiffer than aluminium alloys.

When the stress is a compression, the **bulk modulus** is given by:

Buk modulus = Pressure/Bulk strain

For many engineering materials the tensile modulus of elasticity has the same value as the bulk modulus.

When the stress is a shearing stress, the **rigidity modulus** is given by:

Rigidity modulus = Shear stress/Shear strain

1.6.1 YOUNG'S MODULUS

The results of the tensile test are presented as a plot of stress against strain [see Figure 1.6A].

Stress = Force/Original cross-sectional area: $\sigma = F/A$

Tensile strain = Extension/Original length: $\varepsilon = \Delta l/l$

The value of Young's modulus, the tensile modulus, E, measures the resistance of a material to elastic deformation, that is the **stiffness** of the material.

Young's modulus (tensile modulus) = Stress/Strain

$$E = \sigma/\varepsilon$$

The maximum tensile stress which a material can withstand is its **ultimate tensile stress** or **tensile strength** [see Figure 1.5B].

A sample calculation of the value of tensile modulus

Worked example   A force of 4.00×10^4 N causes an elastic deformation of 0.0750 cm in a 80.0 cm length of alloy rod of diameter 1.50 cm. Calculate the value of Young's modulus for the alloy.

$\sigma = F/A = 4.00 \times 10^4\,N/\pi\,(0.750 \times 10^{-2}\,m)^2 = 2.27 \times 10^8\,N\,m^{-2}$

$\varepsilon = \Delta l/l = 0.0750\,cm/80.0\,cm = 9.38 \times 10^{-4}$

$E = \sigma/\varepsilon = 2.27 \times 10^8\,N\,m^{-2}/9.38 \times 10^{-4} = 2.41 \times 10^{11}\,N\,m^{-2} = 240\,GPa$

Young's modulus = 240 GPa

The values of Young's modulus and tensile strength for some materials are shown in Table 1.6A. Tensile strength and the tensile modulus decrease with increasing temperature. The reason will be covered when the structure of metals has been studied [see § 2.7].

Stress–strain relationships for some different materials are shown.

Materials	*Young's modulus*/GPa	*Tensile strength*/MPa
Mild steel	220	400
Cast iron	150	140–300
Brass	120	120–400
Aluminium alloy	70	150–600

TABLE 1.6A
Young's Modulus
(Tensile Modulus)

The stress–strain relationship for some materials is shown in Figure 1.6A. Polymers have a lower tensile modulus than metals or ceramics. They often undergo a much bigger plastic deformation on loading.

FIGURE 1.6A
Stress–Strain Graphs for a
Number of Materials
(Note the displacement of
the origins)

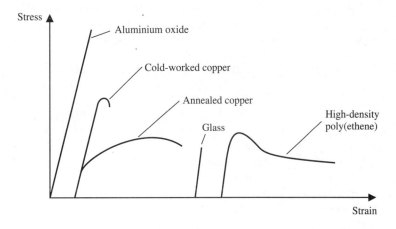

═══════════════════════════ **CHECKPOINT 1.6** ═══════════════════════════

1. Materials **A, B** and **C** have the following values of the tensile modulus (Young's modulus): **A** 1.3 GPa, **B** 2.5 GPa, **C** 3.1 GPa. Which material is the stiffest?

2. A stress of 5.0 MPa produces a strain of 0.020% in a certain material. Calculate Young's modulus (tensile modulus) for the material.

3. A material with tensile modulus 5.0 GPa is subjected to a stress of 5.0 MPa. What strain results?

4. From the stress–strain graph for cast iron, find (a) the modulus of elasticity, (b) the tensile strength of the sample.

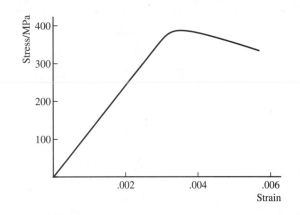

1.7 HARDNESS

A hard material is one whose shape is difficult to change. A hard material will resist abrasion (it wears well) and indentation (it is difficult to dent or scratch). A hard material will dent or scratch a softer material and will withstand the impact of softer materials without changing its shape. Early in the nineteenth century, the German mineralogist Friedrich Mohs defined a scale of hardness, which is shown in Table 1.7A. Each member of the scale is capable of scratching a member with a lower number. Any other mineral can be assigned a number by carrying out a scratch test, for instance a mineral that is scratched by corundum and not by topaz has Mohs hardness 8–9.

10	Diamond
9	Corundum (aluminium oxide)
8	Topaz (aluminium fluoride hydroxide silicate)
7	Quartz (silicon(IV) oxide)
6	Orthoclase (aluminium potassium silicate)
5	Calcium fluoroapatite (calcium chloride fluoride hydroxide phosphate)
4	Fluorite (calcium fluoride)
3	Calcite (crystalline calcium carbonate)
2	Gypsum (calcium sulphate-2-water)
1	Talc (hydrated magnesium silicate)

TABLE 1.7A
Mohs Scale of Hardness

The relative hardness of some materials is shown in Table 1.7B.

Material	Relative hardness	Uses
Diamond	10.0	Jewellery, cutting tools
Silicon carbide	9.7	Abrasives
Tungsten carbide	8.5	Drills
Steel	7–5	Machinery, vehicles, buildings
Sand	7.0	Abrasives, e.g. sandpaper
Glass	5.5	Cut glass can be made by cutting glass with harder materials
Nickel	5.5	Used in coins; hard-wearing
Concrete	4–5	Building material
Wood	1–3	Construction; furniture etc.
Tin	1.5	Plating steel food cans

TABLE 1.7B
Relative Hardness of some Materials

1.7.1 HARDNESS TESTS

The hardness of a solid is assessed by its resistance to indentation, e.g. denting and scratching. Hardness tests for metals involve pressing a hard indenter into the metal surface with a known load and measuring the size of the indentation produced. The **Brinell hardness test** uses a sphere of hard steel of diameter 1–10 mm under a force of 500–3000 kgf (depending on the material) applied for about 20 seconds. (One kilogram force = 9.8 newtons; 1 kgf = 9.8 N.) After the indenter is withdrawn, the diameter of the indentation is measured under a microscope, and the area of the indentation is calculated [see Figure 1.7A].

FIGURE 1.7A
The Brinell Hardness Test

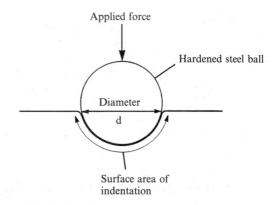

Brinell hardness number, HB = Applied force/Curved area of indentation

The unit is kgf mm^{-2}. The size of the indentation depends on the load and on the diameter of the steel ball. Balls of diameter 1 mm, 2 mm, 5 mm and 10 mm are used. The value of the applied force is regulated so that Force/(Diameter of ball)^2 has the value 1 (for 1 mm ball), 5 (for 2 mm ball), 10 (for 5 mm ball) or 30 (for 10 mm ball); then the same value of hardness number is obtained whichever diameter of ball is used.

Methods of testing hardness by measuring the indentation produced in a specimen by a load include . . .

. . . the Brinell hardness test . . .

. . . and the Vickers hardness test

The **Vickers hardness test** uses a diamond indenter which is a square-based pyramid in shape. A force of 5–120 kgf is applied for 15 seconds, and the size of the impression is measured under a microscope. The Vickers hardness number, HV, and the Brinell hardness number, HB, are approximately equal.

When the hardness of rubber-like materials is measured by a test of indentation, since the materials recover almost completely when the indenter is removed, the indentation is measured while the load is still applied.

1.7.2 IMPACT TESTS

Impact tests measure the response of a material to a sudden high rate of loading. In both the Izod test and the Charpy tests, a heavy pendulum swings through an arc and strikes a notched test piece [see Figure 1.7B].

FIGURE 1.7B Impact Testing – The Izod Test

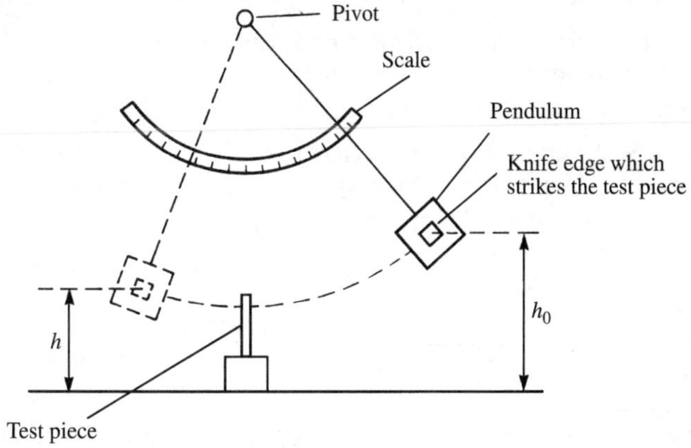

Impact tests measure the response of a material to sudden loading, e.g. the Izod test

After breaking the test piece, the pendulum continues its swing, but some energy has been used to break the test piece and the pendulum does not swing up to the same height as that from which is started. The height to which the pendulum swings is a measure of the energy needed to break the test piece. The energy lost is $mgh_0 - mgh$, where m is the mass of the pendulum, g the acceleration due to gravity, h_0 the starting height of the pendulum and h its maximum height after breaking the test piece. This is called the impact energy. The impact energy is quoted for standard size test pieces and forms of notch.

1.7.3 HARDNESS AND MOLECULAR STRUCTURE

The hardness of a material is related to its molecular structure.

Minerals in which chains or sheets of atoms are held together by van der Waals forces are relatively weak. Talc is a silicate which consists of sheets of covalently bonded silicon and oxygen atoms [see §§ 3.5.2 and 3.5.3]. The bonds between the atoms within the sheet are strong, but the sheets are held together by van der Waals forces and can slide over one another. Talc, which is at the bottom of the Mohs scale, is used as a lubricant. Graphite also has a layer structure with strong bonds between atoms within layers, but only van der Waals forces between the layers. Graphite is a soft substance, which is used as a lubricant [see *ALC*, § 6.6 and Figure 6.15].

In ionic crystals the bonding is strong, but the ionic bonds are non-directional, making deformation not especially difficult. Examples are sodium chloride, with Mohs hardness 2 and calcium fluoride with Mohs hardness 4.

Layer structures are soft. Covalent macromolecular structures are hard.

In substances which consist of covalent macromolecules, the bond strength is greater and the bonding is strongly directional. Diamond, with Mohs hardness 10, has a macromolecular structure [see *ALC*, § 6.5 and Figure 6.11]. Its hardness finds it many uses; cutting and engraving metals and glass, edging saws that can cut through masonry and tipping drilling bits that can drill through rock. Other hard macro-molecular substances are silicon carbide, boron nitride and silicon(IV) oxide, quartz [Figure 3.5F].

The properties of rubbers depend on their structures.

With rubber-like materials, the elastic properties depend on the chain length and the degree of cross-linking between molecules. Short chains are associated with low modulus and low hardness; long chains with a higher modulus and greater hardness [see § 4.19]. The hardest substances are those with a high degree of cross-linking between molecules, e.g. ebonite (a hard black insulating material made by extensive vulcanisation of rubber, that is, treating it with sulphur to achieve cross-linking).

CHECKPOINT 1.7

1. Describe the basic principles of hardness measurement.

2. A plastic has been modified by the incorporation of glass fibres. How can you test to see whether the new material is

(a) harder,   (b) stiffer than the original plastic?

3. (a) What is meant by the statement that a steel has an Izod impact value of 30 J for a 10 mm × 10 mm specimen at 25 °C?

(b) The Izod impact energies for an alloy at different temperatures are tabulated below.

Temperature/°C	+25	−50	−100
Impact energy/J	70	75	85

What can you deduce from these measurements?

4. Refer to the Mohs hardness table. Explain why

(a) diamond is at the top of the scale,

(b) quartz (silicon(IV) oxide) is higher up the scale than calcite (calcium carbonate),

(c) gypsum (calcium sulphate-2-water) is above talc (magnesium silicate hydrate).

1.8 FAILURE

A material is selected for a task on the basis of the properties of the material and an analysis of the task. If the correct material has been chosen for the task, and if the stress levels it will encounter have been correctly calculated, it should not fail in service. A component fails when it no longer fulfils its function, e.g. a piston inside a cylinder block fails if it becomes distorted and jams. Causes of failure include:

- creep [§ 1.9]
- defects in manufacturing, e.g. cracks [§ 1.10]
- loss of ductility through a change in the crystalline state of the material [§ 1.11]
- a sudden impact [§ 1.10.3]
- fatigue [§ 1.12]
- corrosion [§ 2.12]

Failure means that a component can no longer fulfil its function. Some causes of failure are listed.

- Some components are designed to fail under certain conditions in order to protect some other more expensive component, e.g. fuses in electrical components and the use of zinc as sacrificial anodes to protect ships, oil rigs, etc. from corrosion [see § 2.13].

1.9 CREEP

A piece of material may be exposed to a stress for a long period of time. This is a different situation from the tensile test where stresses are applied for short intervals of time. When a stress is applied to a material and allowed to remain acting on the material for a long time, the result is different. The strain increases with time; that is, the material lengthens with time, even though the stress remains constant. The phenomenon is called **creep** It is the continuing deformation of a material with the passage of time when the material is subjected to a constant stress. The increase of creep with time has the form shown in Figure 1.9A. For metals, apart from very soft metals, e.g. lead, creep effects are negligible at ordinary temperatures but become significant at high temperatures. For plastics creep is often quite significant at ordinary temperatures and even more pronounced at higher temperatures. The creep behaviour increases with temperature and increases with stress. Flexible plastics show more creep than stiff plastics.

When stress is applied for a long period of time, the strain may increase with time even though the stress remains constant. This phenomenon is called creep, and can be pronounced in plastics.

FIGURE 1.9A Increase of Creep with Time

Creep increases with temperature . . .
. . . and with the magnitude of the stress . . .
. . . with the composition of alloys . . .
. . . and depends on grain size.

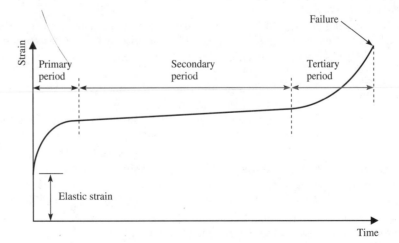

When a metal is selected for a task, creep strain must not be so great as to prevent the metal from functioning under the expected conditions of stress and temperature. The acceptable creep strain is different for different components working in different conditions. For a component which has to operate in a confined space a change in shape is more serious than for one that operates in a free space.

1.10   DEFECTS IN STRUCTURE

1.10.1   CRACKS

Cracks can be a problem in wooden squash racquets. When accidentally smashed against the wall of the squash court, a racquet may break into flying splinters which can cause injury. A solution to the problem of cracks is to employ a composite material [see § 1.2 and Chapter 5]. Carbon-fibre-reinforced epoxy resin is a composite material, consisting of carbon fibres in a matrix of the resin. The fibres stop cracks from spreading through the material by allowing the cracks to spread between the fibres and the epoxy resin [see Figure 1.10A]. Racquets made from this material can absorb impacts without splintering and are therefore much safer.

Cracks are present in all materials . . . . . . and propogate under stress.

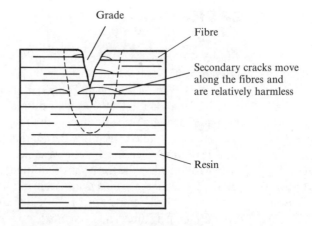

FIGURE 1.10A

A growing crack in a fibre composite material is prevented from spreading by the fibres

The condition for a crack to propagate through a material can be calculated from the properties of the material and the length of the crack.

Cracks are a serious problem when they develop in metal structures such as planes and bridges [see § 1.1].

In 1920, A. A. Griffith put forward the theory that all materials contain small cracks but that a crack will not propagate until a certain stress is reached. The value of the stress needed depends on the nature of the material and the length of the crack. When a crack propagates, it increases the surface area of the solid. To do this, bonds must be broken, and this requires energy. As the crack grows, it allows the material near the crack to relax, and energy (elastic strain energy) is released. When the crack reaches a certain critical size, the released elastic strain energy is just sufficient to provide the energy necessary to break bonds and create new surfaces. The crack can then expand through the material at high speed. In brittle materials, including engineering ceramics, once a minute crack, less than 1 mm long, has formed, if it is subjected to a certain stress it can expand through the material in a fraction of a second. For a ceramic to be strong enough for use as an engineering material the length of the crack must be very small – a few micrometres only. A material with a high resistance to fracture is a **tough material**.

Calculation shows that, for a ceramic, only if a crack is very small, a few µm long, is the ceramic strong enough for engineering use.

1.10.2   STRESS CONCENTRATION

If you want to break a material, you can do so more easily by making a notch in the surface and then applying a force. The presence of the notch makes a big change in the stress at which fracture occurs: it is described as a **stress concentration**. It disturbs the distribution of stress and produces local concentrations of stress. The greater the depth of the notch and the more pointed the tip of the notch, the greater is the local stress.

Stress concentration is high at the end of a notch.

A crack in a ductile material is less likely to lead to fracture than in a crack in a brittle material. A high stress concentration in a notch in a ductile material will deform the notch, increasing the radius of the tip of the notch and consequently decreasing the stress concentration.

1.10.3   SUDDEN IMPACT

A sudden impact may cause fracture where a slow application of the same stress would not.

A sharp blow to a material may cause a fracture where the same amount of stress applied more slowly would not. A fast application of stress may not leave time for plastic deformation of the material to occur and a material which is normally ductile may behave in a brittle manner. If the material has a notch, the likelihood of fracture is still greater.

1.11   TYPES OF FRACTURE

When a ductile material has a gradually increasing stress applied to it, it behaves elastically up to a limiting stress [see Figure 1.4B]. Beyond that plastic deformation occurs. In a case of tensile stress, this takes the form of necking [see Figure 1.11A]. As the stress is increased, the cross-sectional area is further reduced until at some stage fracture occurs. This type of fracture is called **ductile fracture**. The surfaces of the fractured material are dull. Compression, as well as tension, can result in ductile fracture.

FIGURE 1.11A
Ductile Fracture as a Result of (a) Tension, (b) Compression

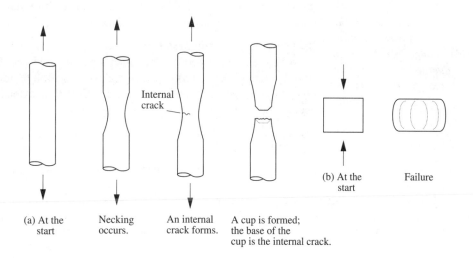

Internal crack

(b) At the start   Failure

(a) At the start   Necking occurs.   An internal crack forms.   A cup is formed; the base of the cup is the internal crack.

A ductile material which is subject to stress undergoes plastic deformation, followed by necking and then by ductile fracture. A brittle material undergoes brittle fracture before much plastic deformation has occurred.

Another type of fracture is **brittle fracture**. This is what happens if you drop a china cup. In brittle fracture the material fractures before any significant plastic deformation has occurred. The fractured surfaces appear bright and granular as they reflect light. This is because fracture has taken place by the separation of grains along specific crystal planes, called cleavage planes [Figure 1.11B].

FIGURE 1.11B
Brittle Fracture as a
Results of (a) Tension,
(b) Compression

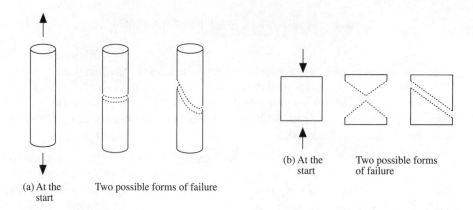

(a) At the start

Two possible forms of failure

(b) At the start

Two possible forms of failure

The temperature of a material can alter its behaviour when subjected to stress. Many metals which are ductile at high temperature become brittle at low temperature [see Figure 1.11C].

FIGURE 1.11C
A Ductile to Brittle
Transition

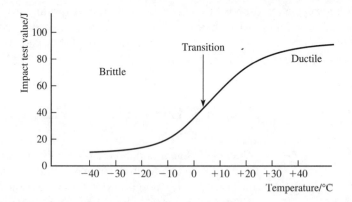

The transition temperature at which the change from ductile to brittle behaviour occurs is one factor that determines how a material will behave in use. At room temperature and above steel is ductile; below 0 °C it behaves as a brittle material. The presence of manganese and nickel lower the transition temperature of steel; the presence of carbon, nitrogen and phosphorus raise the transition temperature.

Some materials change from brittle to ductile behaviour as the temperature rises. The change from brittle to ductile behaviour takes place at the transition temperature.

Brittle fracture is a common form of failure in polymers below the glass transition temperature (the temperature at which a polymer changes from a rigid material into a flexible material; see §4.5). Below the glass transition temperature the molecules are unable to change their configuration; therefore little extension is possible and failure occurs under stress.

═══════════ **CHECKPOINT 1.11** ═══════════

1. (*a*) Distinguish between ductile and brittle fractures.

(*b*) If you were given a fractured specimen, what would you look for to decide whether the fracture was a ductile or a brittle fracture?

2. How does the presence of a notch affect the failure of a specimen of material?

3. Explain why the results of an impact test on steel show a transition from ductile behaviour to brittle behaviour when the temperature is changed.

4. (*a*) Explain what is meant by 'creep'.

(*b*) Describe the form of a typical strain–time graph which results from a creep test

(*c*) Describe the effect of (i) increased stress, (ii) increased temperature on the creep behaviour of materials.

1.12   FATIGUE

While carrying out its function, a component may be subjected to thousands or millions of changes of stress. Some components are repeatedly stressed and unstressed; others undergo alternating compression and tension. Many materials fail under such conditions, even though the maximum stress in any one stress change is less than the fracture stress as measured in the tensile test. Failure which is the result of repeated stressing is called **fatigue failure**. Fatigue is the cause of about 80% of the failures of modern engineering components. The changes in the difference in pressure between the cabin of an aircraft and the outside, which occur each time the aircraft flies, subject the cabin skin to repeated stressing. Turbine blades are subjected to constant vibration.

Fatigue generally starts with the formation of a surface crack, which gradually extends until the peak load can no longer be carried by the reduced cross-sectional area [see § 1.11], and failure occurs rapidly. A fatigue crack often starts at a point of **stress concentration** [see § 1.10.2].

The causes of fatigue include corrosion, overstressing and accidental damage, all of which cause localised damage and a concentration of stress. Fatigue is a major cause of engineering failures, but the reason for failure is the *initial cause* of the stress concentration which in the end leads to fatigue failure.

Cracks always exist in a material as a result of the processing. Should a crack propagate under the conditions of service, it will lead to failure. Metallurgists calculate the size of the crack which will *not* propagate under service conditions [see § 1.10] and then use a non-destructive test to ensure that the only cracks present are smaller than this. The calculations are complicated because many components are subject to a complex stress system. A wide margin of safety must be allowed. Also components of a complex shape may not be suited to the methods available for non-destructive testing.

1.12.1   FATIGUE TESTING

Fatigue testing is carried out by attaching the component to be tested to a machine which applies a repeated stress, bending or twisting or compressing the component until it fails. The machine records the number of cycles of stressing up to failure. The test is repeated at different values of stress. Figure 1.12A shows a plot of a set of results obtained in this way.

FIGURE 1.12A
Results of Fatigue Testing: (a) Steel, (b) a Non-Ferrous Alloy

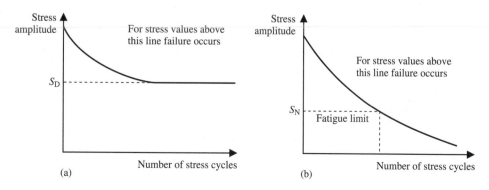

There are two cases:
1. A fatigue limit can be quoted – the minimum stress amplitude at which failure will occur after a sufficient number of cycles.
2. There is no stress amplitude below which failure can never occur. A failure limit after a certain number of cycles is quoted.

In steel, shown in Figure 1.12A (a), at sufficiently low values of stress amplitude, the material will endure an infinite number of stress cycles. There comes a limit: for a stress amplitude S_D, failure will occur after a certain number of stress cycles. The value, S_D is called the **fatigue limit** or **endurance limit**. In Figure 1.12A (b) there is no stress amplitude at which failure *cannot* occur, and a **fatigue limit for N cycles, S_N,** is quoted. For example, for a certain aluminium alloy at a stress amplitude of 185 MPa one million cycles are needed before failure occurs; with a stress amplitude of 115 MPa, one hundred million cycles are needed. If a component made from this alloy has a service life of 100 million stress cycles, one could specify that during its lifetime it should not fail for a stress amplitude below 115 MPa.

CHECKPOINT 1.12

1. Explain what is meant by 'fatigue failure'.

2. List four possible causes of fatigue failure.

3. Cracks are one cause of fatigue failure. What action is taken by metallurgists and engineers to reduce the risk of failure through cracks?

4. Is the quality of being resistant to fracture called (*a*) creep, (*b*) tensile strength, (*c*) toughness, (*d*) ductility or (*e*) hardness?

1.13 ELECTRICAL PROPERTIES

1.13.1 ELECTRICAL CONDUCTORS

Metals are good electrical conductors, whereas polymers and ceramics are generally insulators. **Energy bands** are used to explain the conduction of electricity. The electrons in the shell of an isolated atom have discrete energy levels [see *ALC*, § 2.2]. Those in the highest energy level are the valence electrons, which are used in bond formation [see *ALC*, § 4.2]. When atoms are present in a solid structure, the vibration of the atoms about their mean positions causes the energy levels to spread out into narrow bands. The band containing the valence electrons is called the **valence band**. In metals, electrons in the valence band can move easily to an unfilled energy level in a band of higher energy. This band is called the **conduction band**. Electrons in the conduction band are delocalised; they have broken free from individual atoms and are able to move through the material when a potential difference is applied across it [see § 2.3].

The conduction of electricity is discussed in terms of energy bands . . .
. . . the valence band . . .
. . . the conduction band . . .

In metals there is no energy gap between the valence band and the conduction band. In insulators [see § 1.13.2], the energy gap is too great (about 5 eV) to allow electrons to move between the bands. In semiconductors [see § 1.13.3], the energy gap is smaller (about 1 eV), and the chance of electrons jumping the gap increases as the temperature rises [see Figure 1.13A].

. . . and the relationship between them in metals . . .
. . . in semiconductors . . .
. . . and in insulators.

Materials which are good conductors of electricity have low values of **electrical resistivity** and high values of **electrical conductivity**.

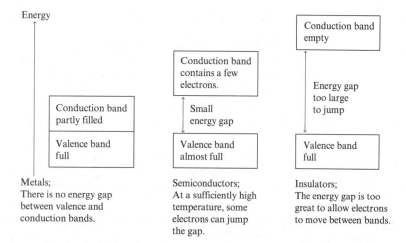

FIGURE 1.13A
Valence Bands and
Conduction Bands

ELECTRICAL RESISTIVITY

The **resistance** of an electrical conductor is given by

$$R = V/I$$

where V = potential difference across the conductor, I = current flowing through the conductor, and R = resistance. The resistance of a conductor depends on its length and cross-sectional area and on the **electrical resistivity** of the material of which it is composed. Electrical resistivity is defined by the equation:

$$\text{Resistivity} = \frac{\text{Resistance} \times \text{Cross-sectional area}}{\text{Length}}$$

Electrical resistivity is defined; its unit is ohm metre.

The unit of resistivity is **ohm metre**, $\Omega\,\text{m}$. (Ω = ohm, m = metre). An electrical conductor, e.g. copper, has a value of the order of $10^{-8}\,\Omega\,\text{m}$, while an insulator, e.g. a ceramic, has a value of the order of $10^{10}\,\Omega\,\text{m}$.

ELECTRICAL CONDUCTIVITY

The **conductance** of a conductor is the reciprocal of its resistance, and **electrical conductivity** is the reciprocal of resistivity:

$$\text{Conductance} = 1/\text{Resistance}$$

$$\text{Conductivity} = 1/\text{Resistivity}$$

Electrical conductivity is defined by the equation:

$$\text{Electrical conductivity} = \frac{\text{Conductance} \times \text{Length}}{\text{Cross-sectional area}}$$

Electrical conductivity is defined; its unit is siemen per metre.

The unit is **reciprocal ohm per metre**, $\Omega^{-1}\,\text{m}^{-1}$, or **siemen per metre**. $\text{S}\,\text{m}^{-1}$, where S = siemen (1 siemen = 1 ohm^{-1}).

Two beautiful conductors

Silver is a good electrical conductor. Small discs of silver make contact in electrical switches in telephones, computers, dishwashers and other electrical appliances. Silver contacts open and close with little friction and therefore little heat is generated.

Silver suffers from tarnishing, especially in polluted air, reacting to form a surface layer of black silver sulphide. For applications where nothing must ever go wrong, such as the electrical circuits in space capsules, gold is used as a conductor because it never corrodes or tarnishes. The connections from silicon chips to electronic circuits are made by fine gold wires.

1.13.2 ELECTRICAL INSULATORS

Non-conductors of electricity or **electrical insulators** are chiefly materials in which the bonding is either ionic or covalent. The valency electrons are retained in their orbitals and cannot move freely as they do in metals. There is a wide energy gap between the valency band and the next possible energy band, which means that a large amount of energy would be required to make an electron cross the gap. A very small number of electrons may acquire that energy if the material is subjected to a high potential difference, and these insulators possess low but measurable conductivities. At high temperatures the electrons have more energy and the chances of an electron escaping increase; that is, conductivity increases. Insulators may break down (become conductors) under a high electrical potential difference if this is high enough to raise the energies of large numbers of electrons to the level where they can cross the gap and become free to move independently. Such a potential difference is called the **breakdown voltage**.

Electrical insulators have ionic or covalent bonding. Insulators become conducting above their breakdown voltage.

1.13.3 SEMICONDUCTORS

Semiconductors are a special class of materials which have an electrical resistivity between those of electrical conductors and electrical insulators. They all conduct electricity better as the temperature rises; that is, the resistivity falls as the temperature rises.

Semiconductors include both elements and compounds. The semiconductors which are pure elements and compounds are described as **intrinsic semiconductors**. Silicon is the intrinsic semiconductor which is used for the production of the integrated circuits known as **silicon chips**. At absolute zero, all the electrons are localised in the valence band, and the material is an electrical insulator. When heat is supplied to a piece of silicon, some of the electrons gain enough energy to jump from the valence band into the conduction band. These delocalised electrons can carry charge through the material when a potential difference is applied across it. As the temperature rises, more electrons jump into the conduction band, and the resistance of the material falls.

Semiconductors are introduced . . .
. . . both intrinsic,
e.g. silicon . . .
. . . and extrinsic.

Extrinsic semiconductors are made by adding certain substances to intrinsic semiconductors. The process is called **doping**, and the added substance are called **dopants**. The dopants are chosen to produce n-type semiconductors, which have a negative charge or p-type semiconductors which have a positive charge. For more information about semiconductors, see *ALC*, § 23.4.

1.13.4 SUPERCONDUCTORS

FIGURE 1.13B
A Body Scanner

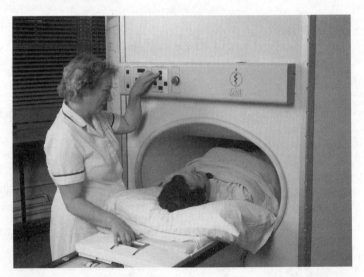

The body scanner shown in the photograph uses a very strong magnetic field which makes atoms in the body emit tiny radio signals. The signals are picked up by a detector and analysed by a computer which is linked to a screen which displays an image of a cross-section of part of the patient's body. The patient feels no pain and suffers no damage.

A magnetic field is generated when an electric current passes through a coil of wire (a solenoid). Heat is generated, and the size of current that can be passed is limited by the rise in temperature that the coil can withstand. Materials called **superconductors** have zero resistance and can pass huge currents without becoming hot. When a solenoid is wound with a superconductor, it can take a large current and can give a very strong magnetic field. The use of a superconductor makes it possible to generate the high magnetic field needed for a body scanner. The first superconductors had to be cooled in liquid helium. With its boiling temperature of 4.2 K, liquid helium is both costly and difficult to work with because of the insulation required.

A major breakthrough came in 1986 when two Swiss scientists, Alex Müller and Georg Bednorz, discovered some ceramics which are 'high temperature' superconductors. They included a ceramic (one containing copper, oxygen, barium and yttrium) with a superconducting temperature of 93 K. Although this may not sound a 'high temperature' by any other standard, it broke the 'liquid nitrogen barrier', 77 K; that is, it would superconduct when cooled in liquid nitrogen. The use of liquid nitrogen, which costs about the same as petrol and requires less insulation than liquid helium, opened the door to superconductor applications. Müller and Bednorz won the Nobel Prize in 1987. Now the race is on to produce a material with room temperature superconductivity!

Superconductors have zero resistance and can allow the passage of huge currents. They are used to produce strong magnetic fields.

When a material is heated, it may absorb heat and rise in temperature, it may expand, and it may transmit heat. These properties can be measured as follows.

1.14 THERMAL PROPERTIES

When a material is heated, it may absorb heat and rise in temperature, it may expand, and it may transmit heat. These properties can be measured as follows.

1.14.1 SPECIFIC HEAT CAPACITY

Specific heat capacity measures the quantity of heat needed to raise the temperature of unit mass of material by one degree. It is defined by:

$$\text{Specific heat capacity} = \frac{\text{Quantity of heat}}{\text{Mass} \times \text{Change in temperature}}$$

Specific heat capacity is defined: its unit is joule per kilogram per kelvin.

The unit is **joule per kilogram per kelvin**, $J\,kg^{-1}\,K^{-1}$.

1.14.2 THERMAL EXPANSION

The **linear expansivity** is a measure of the rate at which a material will expand when the temperature rises. It is defined by the equation:

$$\text{Linear expansivity} = \frac{\text{Change in length}}{\text{Original length} \times \text{Change in temperature}}$$

The unit is **reciprocal kelvin**, K^{-1}.

Linear expansivity is defined: its unit is reciprocal kelvin.

Polymers have high linear expansivity, and an increase in temperature can cause a big change in the dimensions of polymers. The expansion of polymers, unlike metals, is not usually linear with temperature, and the linear expansivity usually increases with temperature.

1.14.3 THERMAL CONDUCTIVITY

The **thermal conductivity** of a material is a measure of the rate at which heat will flow through the material. Thermal conductivity is defined by the equation:

$$\text{Thermal conductivity} = \text{Rate of heat flow}/(\text{Area} \times \text{Temperature gradient})$$

The unit of thermal conductivity is **watt per kelvin per metre**, $W\,K^{-1}\,m^{-1}$.

Thermal conductivity is defined; its unit is watt per kelvin per metre.

Metals have high thermal conductivity; polymers and ceramics have much lower values. If heat insulation is required, polymers and ceramics are feasible; if good heat conduction is required, metals are appropriate.

1.15 DENSITY

Density is defined; its unit is kilogram per cubic metre . . .
. . . or kilogram per cubic decimetre.

The **density** of a material is given by the equation:

$$\text{Density} = \text{Mass}/\text{Volume}$$

Density has the unit kilogram per cubic metre, $kg\,m^{-3}$, or kilogram per cubic decimetre, $kg\,dm^{-3}$.

Examples are given of the importance of density in the choice of materials.

Density is often an important factor in the choice of a material. For aircraft manufacture, a material of low density and high strength, e.g. an aluminium alloy, has an obvious advantage. When choosing a material from which to make soft-drink bottles, manufacturers consider the densities of rival materials. The low densities of plastics mean that a crate of plastic bottles of a drink will weigh less than a crate of glass bottles of drink and transport costs will be lower.

Who wants a high-density metal?

Sometimes a high density is an advantage. Tungsten has a density 2.5 times that of iron. It is also extremely hard. This combination suits the darts player. In order to score a treble, the player has to get three darts in a small area. Darts of 26–28 grams are the easiest to throw accurately. Darts made from tungsten and nickel are only half as thick as the traditional brass darts and therefore enable the expert to make high scores. They cost many times more than brass darts.

CHECKPOINT 1.15

1. Copper has a resistivity of $1.6 \times 10^{-5}\,\Omega\,m$. Calculate the resistance of a copper wire of length 1 m and cross-sectional area $2\,mm^2$.

2. The dielectric strength of poly(ethene) is $4 \times 10^7\,V\,m^{-1}$. What voltage is required to overcome the insulation of a piece of poly(ethene) of thickness 1 mm?

3. Suggest the main property or properties required of materials that could be used in the following applications:

(*a*) a component which must be strong at a temperature of 700 °C,

(*b*) a component which must be stiff,

(*c*) a component which must be strong and which is to be used in a marine environment,

(*d*) a heating element to be used at temperatures of over 1000 °C,

(*e*) a component subject to impact loading,

(*f*) a component subject to cyclic loading,

(*g*) an electric power cable.

QUESTIONS ON CHAPTER 1

1. Which properties of a material describe the following?

(*a*) its stiffness, (*b*) its ability to be bent into shape, (*c*) its ability to conduct electricity, (*d*) its ability to conduct heat, (*e*) the difficulty of breaking it, (*f*) the ability to withstand force when cracks are present, (*g*) the ability to withstand abrasion.

2. The strength of a material may be quoted in terms of yield stress, tensile strength or fracture stress. Explain the difference between the three quantities. Comment on their relative usefulness. Explain how these three quantities are related to bonding in metals.

3. Sketch the form of the stress–strain graphs for
(*a*) a strong brittle material, (*b*) a strong ductile material, (*c*) a weak ductile material.

4. From the stress–strain graph for the polymer ABS, find
(*a*) the modulus of elasticity, (*b*) the tensile strength of the specimen.

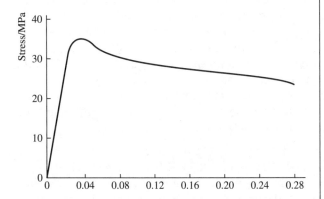

5. The figure shows how the strain changes with time for two different polymers subjected to a constant stress. Say which material will creep the most.

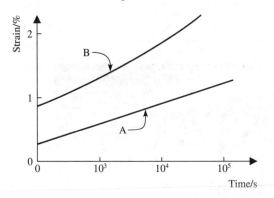

6. Explain what is meant by fatigue failure.

7. What is meant by the statement that a material has
(*a*) a high tensile strength,
(*b*) a high tensile modulus,
(*c*) a low yield stress,
(*d*) a low impact energy?

8. (*a*) What is the tensile stress acting on a metal rod of cross-sectional area $50 \, mm^2$ when the tensile force acting on it is $1000 \, N$?

(*b*) A tensile force causes a rod of length $40 \, cm$ to extend by $2 \, mm$. Calculate the strain.

9. A metal specimen extends as the load applied to it increases as shown in the figure until at point E it fractures.

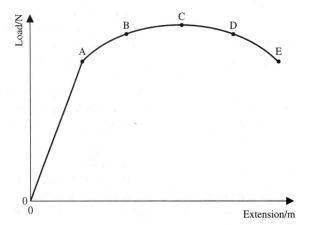

(*a*) What name is given to point A?

(*b*) On a copy of the curve, sketch the curve that results when the specimen is unloaded from point B to zero load.

(*c*) What does point C represent?

(*d*) A similar specimen failed after repeated loading at only half the load experienced at point A. Explain (i) why this could happen and (ii) what steps can be taken to avoid such a failure.

2

METALS

2.1 METALS IN DAILY LIFE

As the shadows lengthen, I need some light on my work. I stretch out my hand to press a switch, and current flows along a copper wire to heat the tungsten filament of an electric light bulb. I shiver slightly and press another switch, which allows current to pass to an electric fire with an element of nickel–chromium alloy. I have driven home from the library in my car, 75% of which is composed of metals and alloys, chiefly steel and aluminium. I fed the parking meter with coins made of a copper–zinc–nickel alloy. When dinner time approaches, I shall use my electric cooker of steel and aluminium, while listening to the radio which contains magnets made of steel alloys with rare metals such as neodymium. I shall turn a brass tap plated with nickel and chromium to allow water to run along a copper pipe into my sink, which, like the saucepans, knives and forks, is made of stainless steel.

Metals play an essential role in our everyday life.

The part that metals play in our daily lives cannot be overestimated. Before the Industrial Revolution of 1780–1860, the people of the UK lived in small self-supporting communities based on agriculture. The development of industry brought about a movement to large towns and cities based on manufacturing. The machines that made the Industrial Revolution possible were made of iron and steel.

Today all our means of transport – motor vehicles, trains, ships, planes – and the bridges which carry our vehicles are made of steel, aluminium alloys and other alloys. The tunnels through which we can travel under the River Mersey and the River Severn and the Channel have been excavated with tungsten alloy drills.

FIGURE 2.1A An Aluminium Alloy Train

2.2 SOME CHARACTERISTICS OF METALS

You will be familiar with the physical and chemical properties of metals from your earlier studies.

1. Metals are solid elements, except for mercury which is liquid at room temperature.

2. They are **lustrous** when a fresh surface is exposed but most of them become tarnished by exposure to the air.

3. Metals are **sonorous**: they make a pleasing sound when struck.

4. They are **ductile**: they can be worked into different shapes without breaking.

5. Metals are good electrical and thermal conductors.

6. The oxides of metallic elements are basic or amphoteric; see *ALC*, § 28.8.1.

Characteristically, metals are lustrous, sonorous, ductile and crystalline ...

7. Metals are **crystalline**. You can see crystals of zinc on the surface of galvanised iron (iron plated with zinc). You can watch crystals of lead form if you immerse a strip of zinc in a solution of lead (II) nitrate.

...with the exception of mercury. Mixtures of metals are called alloys.

For some applications, many metallic elements are too soft. Metallurgists formulate mixtures of metals called **alloys**. Alloys are harder and therefore more useful than the elements from which they are made. The huge number of alloys that can be made puts an enormous variety of metallic materials at our disposal. Examples of aluminium alloys, copper alloys and steel are given in § 2.2 and § 2.4.

2.3 THE METALLIC BOND

The properties of metals derive from the nature of the metallic bond. Metal atoms in a piece of metal have overlapping atomic orbitals ... ... which enable outer shell electrons to become delocalised ... ... and move through the structure at random. The metal cations are held in position by an electrostatic attraction to the electron cloud.

Strong forces of attraction exist between the atoms in a metal. The nature of these metallic bonds is described in *ALC*, § 6.2.1. The outer shell electrons from each atom come under the influence of a very large number of atoms. They are not attached to any particular atom and are described as **delocalised electrons**. The delocalised electrons move at random through the structure. As the outer shell electrons have been removed from metal atoms, metal cations have been formed. The cations are not pushed apart by the repulsion between them because they are attracted to the electron cloud between them. The properties of metals can be explained in terms of the metallic bond; see *ALC*, § 6.2.1 and Figure 6.3.

The strength of the metallic bond depends on the number of electrons which can become delocalised. A Group 2 element with two electrons in the outer shell forms stronger bonds than a Group 1 element with only one electron in the outer shell. Transition metals can use electrons from the d shell in addition to the outer shell and therefore form stronger metallic bonds than the s block metals. This is why transition metals are used for the construction of vehicles, ships, bridges, tools and buildings.

2.4 ALLOYS

An alloy is a metallic substance composed of a metallic element mixed with one or more other elements. Pure metals, with high ductility, low tensile strength and low yield strength, tend to be too soft for many manufacturing and engineering purposes. Alloying can give harder substances, with lower ductility, higher tensile strength and higher yield strength. Most of the metallic objects which you see around you and use daily are made of alloys. There are some purposes for which pure metals are better, e.g. applications which require high electrical conductivity or good corrosion

Pure metals are too soft for many purposes. Alloying can increase their hardness and strength.

resistance or high ductility. Alloys have lower conductivity, corrode more rapidly and are less ductile than the elements of which they are composed. The structure of alloys will be described in § 2.8.

2.4.1 STEELS

The most widely used alloy is steel. There are many types of steel. The **plain carbon steels** are alloys of iron and carbon [see Table 2.4A].

Type of steel	Carbon/%	Uses
Mild steel	0.07–0.25	Easily cold-worked, the most widely used type of steel; car bodies, drawn tubes
Medium carbon steel	0.25–0.55	Wear-resisting; shafts, axles, wires, rails and cylinders
High carbon steel	0.55–0.90	Strong and wear-resistant; hammers, wrenches, chisels, drills
Carbon tool steel	0.90–1.6	Strong and wear-resistant; ball bearings, machine parts, boring and finishing tools
Cast iron	2.5–3.8	Excellent casting properties; automobile cylinders, pistons, moulds, machine castings

Plain carbon steels are alloys of iron and carbon. The table shows some of their uses.

TABLE 2.4A
Plain Carbon Steels

Alloy steels have two types of applications: as structural steels and as stainless steels. Some of the large number of alloy steels are tabulated below [Table 2.4B].

Element	Characteristics and uses of alloy steel
Cobalt	High magnetic permeability; magnets
Manganese	Strong and hard; engineering materials
Molybdenum	Maintains strength at high temperature; gun barrels
Tungsten	High melting temperature, tough; high-speed tools, dies
Titanium	Maintains strength at high temperature; gas turbines, spacecraft
Nickel and chromium together	Stainless steels: resist corrosion by forming an impermeable oxide film; chemical plant construction, turbine blades, rivets, springs, ball bearings, food industry, surgical instruments, cutlery, etc.

Alloy steels contain one or more metals in addition to iron.
There are many alloy steels ...
... of which the table shows a few.

TABLE 2.4B
Some Alloy Steels

2.4.2 COPPER ALLOYS

Copper has high electrical and thermal conductivities, is very ductile and resists corrosion. It is used for electrical wires and cables and for pipes and tubes. Some alloys based on copper are listed in Table 2.4C.

Copper is used in electrical circuits.
Some of its alloys are tabulated.

Alloy	Composition	Uses
Brass	10% zinc 30% zinc	Jewellery Cartridge cases, shell cases, wire, tubes
Muntz metal	40% zinc and other elements	Marine propellers and shafts, autoclaves
Tin bronze	up to 20% tin	Fittings, bearings, reactors for processes involving water
Nickel silver	20–30% nickel	Coinage
Monel metal	70% nickel	Turbine blades, corrosion-resistant machine parts

TABLE 2.4C
Some Alloys of Copper

2.4.3 ALUMINIUM ALLOYS

Aluminium is low in density and its alloys are therefore used in the aircraft industry. It has high thermal and electrical conductivity. Being a better electrical conductor than copper, it is used for power cables (with a steel core for extra strength). It has excellent corrosion resistance due to the oxide layer which forms on the surface and is therefore used in the canning industry and in construction.

FIGURE 2.4A
The Airbus is made of aluminium alloy

Aluminium alloys are low in density and used in aircraft manufacture.
The corrosion-resistance of aluminium is exceptional.
Duralumin is an important alloy.

- The most widely used aluminium alloy is Duralumin, an alloy with copper (5%) and a little magnesium, manganese and silicon.

- The best corrosion resistance is found in an alloy with magnesium and silicon.

- An alloy which has a relatively low melting temperature and can be cast is the eutectic mixture [see § 2.9] of aluminium with 13% of silicon.

- The highest strength is found in an alloy with copper, magnesium and zinc.

2.4.4 SOLDERS

Solders are low melting temperature alloys.

The use of alloys of lead and tin as solders is due to the low melting temperature of the eutectic which the metals form [see § 2.9].

2.4.5 HIGH-TEMPERATURE SERVICE ALLOYS

Alloys which retain their strength at high temperature are important in jets, rockets and nuclear plants.

The development of alloys which will function at high temperatures has become increasingly important in the manufacture of rockets, jets and nuclear plants. Examples are Nimonic 80 A, which is based on nickel with 21% chromium, 3% titanium and other elements, and Ta 7 Cr alloy, which is based on iron with 7% chromium and 1% tantalum.

The aerospace metal

Titanium is sometimes described as the **aerospace metal**. It has a unique set of characteristics:

- a low density, only half that of steel

- a high tensile strength of 300–750 MPa, which alloying increases to 1250 MPa

- an exceptionally high resistance to corrosion by acids, alkalis, sea water, etc. This is due to an oxide layer on the surface of the metal.

- a high melting temperature, 1660 °C, which enables titanium to retain its strength at high temperatures

Titanium has a set of properties which make it uniquely suited to aircraft manufacture ...
... and in other applications where resistance to corrosion is a necessity.

The unique combination of properties makes titanium a good choice of material for high-speed aircraft. It is alloyed with other transition metals to increase its strength. Titanium alloys have high fatigue resistance. They are much more costly than stainless steel. The Blackbird, the fastest aircraft in the world, which can travel at four times the speed of sound, consists of 85% titanium alloy. The Tornado, another military aircraft, contains about 25% titanium alloy. The Boeing 777 and Concorde contain components made from titanium alloys. Several helicopters rely on titanium alloys for rotor heads, blade attachments, etc.

---

1. Our theory of the metallic bond envisages a piece of metal as consisting of cations and delocalised electrons.

(*a*) What is meant by a *delocalised* electron?

(*b*) Why do the cations not repel one another?

(*c*) How does this picture of the metallic bond explain the ductility of metals and their thermal and electrical conductivity?

2. Why do transition metals have high tensile strength and compressive strength?

3. What advantages do alloys have over metallic elements? In what circumstances would a pure metal be chosen over an alloy?

4. Suggest an alloy which is

(*a*) exceptionally strong at high temperature

(*b*) used in cartridge cases

(*c*) used in coinage

(*d*) resistant to chemical attack

(*e*) used in aircraft construction

(*f*) of low melting temperature

2.5 THE CRYSTAL STRUCTURES OF METALS

Metals are crystalline: the atoms pack closely together in a regular structure. It is impossible to pack spheres without leaving spaces between them. Arrangements in which the gaps are kept to a minimum are called **close-packed** structures. X-ray studies have revealed three main types of metallic structures. In the **hexagonal close-packed structure** and the **face-centred-cubic close-packed structure**, the metal atoms pack to occupy 74% of the space. In the **body-centred-cubic structure**, the atoms occupy 68% of the total volume.

The way in which the structures are built up can be envisaged by considering the stacking of spheres. It will help you to envisage the structures if you experiment with the stacking of marbles, table tennis balls or styrofoam spheres. In a layer A of identical spheres packed together on a flat surface, each sphere is in contact with six others [Figure 2.5A (a)]. There are spaces or holes between the spheres. Imagine laying a second layer B on top of the first. Each sphere in layer B will rest in a depression between three spheres in layer A. The holes in layer B are of two kinds [Figure 2.5A (b)]. Tetrahedral holes in layer B lie over spheres in layer A, and octahedral holes in layer B lie over holes in layer A.

There are two ways in which a third layer can be laid on layer B. The spheres of the third layer can cover the tetrahedral holes; in this case the third layer is identical with layer A. This type of structure is made up of alternating layers ABABA and is described as a hexagonal close-packed structure [Figure 2.5A (c)]. Since every atom is in contact with twelve others (six in the same layer, three in the layer above and three in the layer below), it is said to have a **coordination number** of 12.

Alternatively, the spheres in layer C can cover the octahedral holes in layer B so that layer C is not identical with layer A. When a fourth layer of spheres is laid on layer C it is identical with A. This type of structure with the sequence ABCABC is a face-centred-cubic close-packed structure [Figure 2.5A (d)]. The coordination number is 12. The high coordination numbers in these structures arise from the non-directed nature of the metallic bond.

Shown in Figure 2.5A (e) and (f) are the **unit cells** of the two types of structure. A unit cell is the smallest part of the crystal that contains all the characteristics of the structure. The whole structure can be generated by repeating the unit cell in three

Metals are crystalline. The crystal structures are of three main types . . . . . . the hexagonal close-packed structure . . . . . . the face-centred-cubic close-packed structure . . . . . . and the body-centred-cubic structure (which is less close-packed).

The number of neighbouring atoms with which an atom is in contact is called the coordination number.

FIGURE 2.5A

The Packing of Spheres in the Hexagonal Close-Packed and the Face-Centred-Cubic Close-Packed Structures

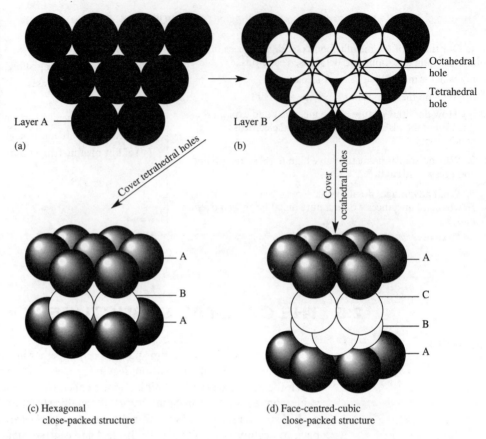

Layer A

(a)

Layer B

Octahedral hole

Tetrahedral hole

(b)

Cover tetrahedral holes

Cover octahedral holes

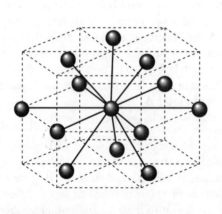

A

B

A

(c) Hexagonal close-packed structure

A

C

B

A

(d) Face-centred-cubic close-packed structure

Values of coordination number are: hexagonal close-packed structure 12 face-centred-cubic close-packed structure 12 body-centred-cubic structure 8

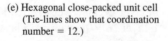

(e) Hexagonal close-packed unit cell (Tie-lines show that coordination number = 12.)

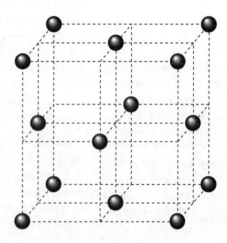

(f) Face-centred-cubic unit cell

Tetrahedral hole

Octahedral hole

dimensions. The less closely packed body-centred-cubic structure is shown in Figure 2.5B. With one atom at each of the eight corners of a cube and one in the centre touching these eight, the coordination number is 8. Figure 2.5B (b) shows an expanded view and Figure 2.5B (c) shows the unit cell with tie-lines to show that the coordination number is 8.

FIGURE 2.5B
Body-Centred-Cubic
Structure

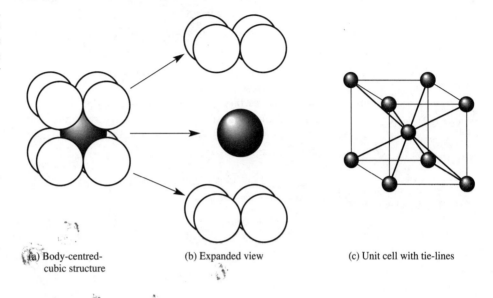

(a) Body-centred-cubic structure   (b) Expanded view   (c) Unit cell with tie-lines

2.6 SINGLE CRYSTALS AND POLYCRYSTALLINE SOLIDS

2.6.1 GRAINS

Metal objects are formed by casting. They may be simple shapes which are subsequently developed or objects in their final form. The casting process is controlled by the temperature, the rate of casting, the mould material and additions to the melt. As a melt cools, small nuclei of the solid appear in the liquid. As cooling continues, more solid deposits on these nuclei to form small crystals. Many metals form crystals of the shape shown in Figure 2.6A and described as **dendrites**. Each nucleus gives rise to a single crystal. The crystals grow until eventually they meet to form a solid mass of small crystals, a **polycrystalline solid**. The individual crystals are called **grains** [Figure 2.6B]. Each grain grows from the liquid metal independently of its neighbours. The

FIGURE 2.6A
A Dendrite

Many molten metals cool to form crystals of dendrite shape.
A piece of metal is a polycrystalline solid ...

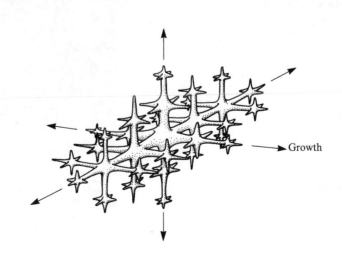

Growth

FIGURE 2.6B
Arrangement of Metal Grains

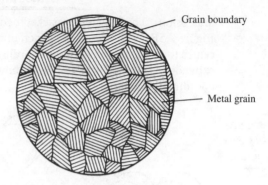

Grain boundary

Metal grain

... consisting of a mass of small individual crystals ...
... which are called grains ...

directions in which the atoms are arranged consequently bear no relationship to those in neighbouring grains [Figure 2.6C]. At the boundaries between the grains, there are a few atoms that have not fitted into the crystal structure of either of the adjacent grains.

FIGURE 2.6C
Boundaries Between Grains

Between the grains are atoms which have not fitted into the crystal structure. These may be atoms of impurities.

Grain boundary

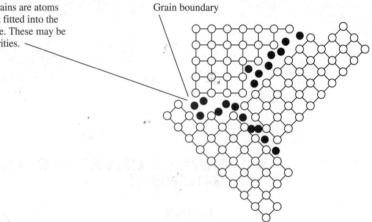

... and are separated by grain boundaries from neighbouring grains.

2.6.2 SINGLE CRYSTALS

It is possible to solidify a molten metal so that it will form a single crystal. The absence of grain boundaries gives superior mechanical properties. The first step in the production of a large single crystal is to select one grain of the metal which is free

FIGURE 2.6D
This turbine blade is a single crystal

Metals can be solidified from the liquid state in such a way that the solid formed is a single crystal.

from defects and has the correct crystalline shape. This is done by X-ray examination. The selected grain, the 'seed', is attached to a holder which is positioned above a container of the molten metal. The seed is lowered until one face touches the metal. The seed is slowly rotated and slowly withdrawn at a rate of a few centimetres per hour. Metal attaches to the seed, which grows into one large single crystal. The process is usually carried out at reduced pressure.

The technology has advanced to the point where large components can be made from single crystals of alloys. Single crystal technology has made it possible to operate jet engines at higher temperatures, with increased fuel economy and increased power.

2.7 DISLOCATIONS IN THE CRYSTAL STRUCTURES OF METALS

2.7.1 SLIP

Metals can change shape in response to stress without fracturing.

How do metals yield to stress without fracturing? The simplest model of the plastic behaviour of metals under stress is the **block slip theory**. When a sufficient stress is applied to a crystal structure, e.g. a metal, blocks of atoms become displaced [Figure 2.7A]. When the yield stress is reached, there is a movement of large blocks of atoms as they slip past each other. The relative sliding of two parts of a crystalline structure on either side of a plane is called **slip**, and the plane is called the **slip plane**.

FIGURE 2.7A Block Slip Model of Plastic Behaviour of Metals

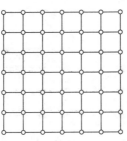

(a) Metal structure

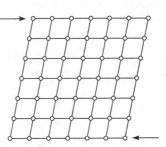

(b) Stress applied to a metal structure results in elastic strain.

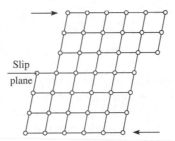

Slip plane

(c) Stress applied results in yielding.

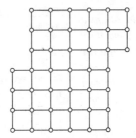

(d) Stress is removed and permanent deformation has occurred.

Blocks of atoms can slide past other blocks of atoms along a slip plane.

2.7.2 DISLOCATIONS

The ductility of metals is much higher than calculated on the basis of the slip model, and the theory must be modified. In the 1950s, the transmission electron microscope came into widespread use. Examination of metal structures showed the presence of defects in the crystal structure. Figure 2.7B shows such a defect. It is as though a half-

FIGURE 2.7B
A Dislocation within a
Metal Grain

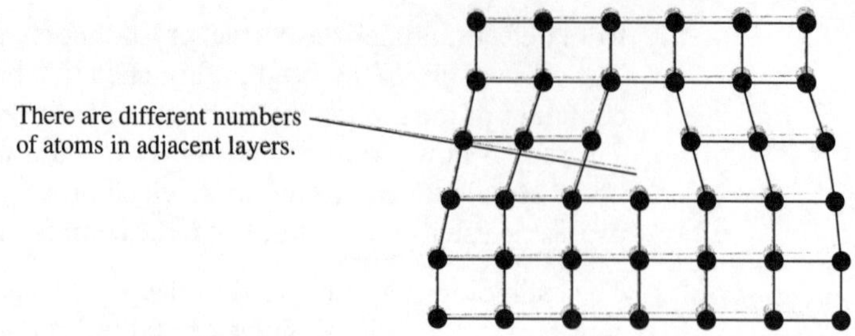

*Metals are much more
ductile than the slip model
predicts.
The high ductility arises
from the presence, in the
crystal structure of the
metal, of defects called
dislocations.*

There are different numbers
of atoms in adjacent layers.

plane of atoms is missing, and the neighbouring planes are slightly displaced to minimise the strain. There are millions of these faults, called **dislocations** in every cubic millimetre of metal.

In an electron microscope, with a magnification of several hundred thousand, dislocations can be seen as thin wavy lines which sometimes move within the crystal. The movement of a large number of dislocations together brings about an irreversible deformation in the crystal.

*Under stress, dislocations
move through the
structure . . .
. . . without affecting the
majority of bonds.
In the movement of a
dislocation, fewer bonds
are broken and reformed
than in the slip of large
blocks of atoms.*

When a metal crystal is stressed, e.g. when a metal is hammered, the structure is distorted and a dislocation may be forced to move. As the stress is repeated, the dislocation may travel along to the grain boundary [Figure 2.7C]. The shape of the grain has been changed. The dislocation moves through the structure without affecting the majority of bonds and without large-scale movement of planes of atoms past each other. This is much easier than slip. The mobility of dislocations explains the ductility of metals.

FIGURE 2.7C
Dislocations Moving
Towards a Grain
Boundary

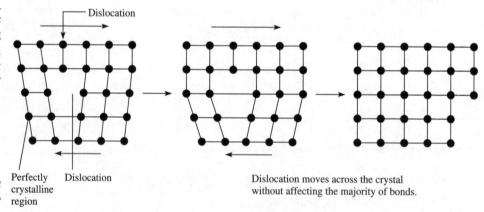

Perfectly
crystalline
region

Dislocation

Dislocation moves across the crystal
without affecting the majority of bonds.

*The more dislocations a
block of metal has the more
likely it is that the
movement of one
dislocation will obstruct
the movement of another.
After repeated stress,
dislocations no longer
move easily; the metal is
work-hardened.
Movement of dislocations
is hindered by grain
boundaries . . .
. . . and by imperfections in
the structure.*

When two dislocations meet, the movement of one dislocation can interfere with the movement of the other and hinder its progress. The more dislocations a metal has, the more the dislocations get into the way of each other and the more difficult it is for dislocations to move through the material. More stress is needed to make the metal yield. After repeated stress the dislocations no longer move easily, and the metal is described as **work-hardened** [see § 1.5.2 and § 2.11.1].

The movement of dislocations is also hindered by grain boundaries. The more grain boundaries there are in a metal, the more difficult it is to make that metal yield, and the stronger is the metal. More grain boundaries occur when grain size is small. A metal which contains a large number of small grains is described as a 'fine-grained' metal. A fine-grained metal is harder than a coarse-grained metal. 'Grain refining agents' can be added to molten metal to provide nuclei from which crystals grow to produce a fine-grained metal.

2.7.3 TEMPERATURE

As the temperature falls, the movement of dislocations becomes more difficult, and the yield stress increases. At low temperature, a transition from ductile behaviour to brittle behaviour occurs.

As the temperature falls, the yield stress increases because it becomes more difficult for dislocations to move through grains. Experiments show that the effect is relatively small in face-centred-cubic metals and alloys, and is marked in body-centred-cubic structures. On impact testing, body-centred-cubic metals, e.g. mild steel, show a sharp transition between ductile and brittle behaviour. The problem is great in the presence of stress-raisers, e.g. notches. The tensile modulus and the tensile stress decrease with temperature [see Figure 1.11C].

═══════════════ CHECKPOINT 2.7 ═══════════════

1. Explain what is meant by (*a*) a single crystal, (*b*) a polycrystalline material.

2. Explain (*a*) what is meant by a dislocation, (*b*) how the malleability of a metal depends on the presence of dislocations.

3. Explain why cold working makes a metal harder.

4. List three ways of increasing the yield stress of a metal. Explain how each method works.

2.8 SOLID SOLUTIONS

Metals are usually soluble in one another in the liquid state and usually immiscible in one another in the solid state.

Usually metals are completely soluble in one another in the liquid state. When the liquid solution is cooled, the two metals separate to form two separate solid phases. The term **phase** is used to describe any part of a system which is homogeneous (having the same physical and chemical properties throughout) and is physically distinct from all the other parts of the system. Lithium and caesium dissolve in each other in all proportions in the liquid state and are completely immiscible in the solid state [Figure 2.8A]. However, some solutions of liquid metals remain as solutions when they solidify, without either metal separating: a **solid solution** is formed. Copper and nickel form a continuous solid solution: they dissolve in one another in all proportions in the solid state [see Figure 2.8B].

FIGURE 2.8A
The Lithium–Caesium
Phase Diagram

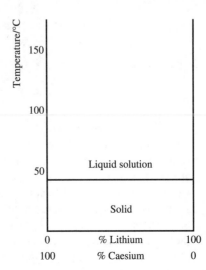

FIGURE 2.8B
The Copper–Nickel Phase
Diagram

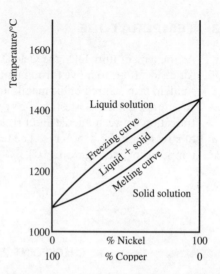

FIGURE 2.8B
The Copper–Nickel Phase
Diagram

It is more likely that a solid solution will be formed when the proportion of an alloying element added to the solvent liquid metal is small. Solid solutions can be **substitutional** or **interstitial**. In substitutional solid solutions, atoms of the solute element occupy some of the positions of atoms of the solvent element in the solid structure [see Figure 2.8C]. In interstitial solid solutions, solute atoms fit into the interstices (spaces) between atoms of the solvent metal [see Figure 2.8D].

There are exceptions which form solid solutions ...

FIGURE 2.8C
Substitutional Solid
Solution

... which can be substitutional when the atoms of the two elements are similar in size ...

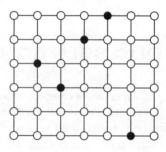

Metal in substitutional solution has little effect on hardness.

FIGURE 2.8D
Interstitial Solid Solution

... or interstitial when there is a big difference in size between the atoms of the two elements.

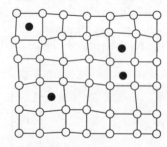

Metal interstitial solution hardens by distorting the lattice thus preventing, the movement of dislocations.

Substitutional solid solutions are formed when the atoms of the two elements are similar in size, e.g. copper and nickel [Figure 2.8B]. The alloys constantan (60% Cu, 40% Ni), which is used in coinage, and Monel metal (60% Cu, 35% Ni, 5% Fe), which is highly resistant to corrosion and is used in pumps and piping, are solid solutions. There is a wide range of alloys formed between transition metals since they have atoms of similar sizes and can form interstitial solid solutions. The presence of the larger atoms of an alloying metal in a metal crystal interferes with the movements of dislocations. This is what makes the alloy stronger, harder and less ductile.

The formation of interstitial solid solutions is limited by the need for a large difference in size between atoms of the two elements. Interstitial solid solutions of carbon in iron are important in the formation of steel [see § 2.10.6].

2.9 EUTECTICS

2.9.1 THE SHAPES OF COOLING CURVES

When a pure liquid cools, the temperature falls until it reaches a value at which solid starts to form. The temperature remains constant at this value until all the liquid has solidified. The reason is that the enthalpy of melting is being evolved. When all the liquid has solidified, the temperature begins to fall again. In practice, supercooling often occurs and the curve is of the shape shown in Figure 2.9A.

FIGURE 2.9A
A Cooling Curve for a
Pure Liquid

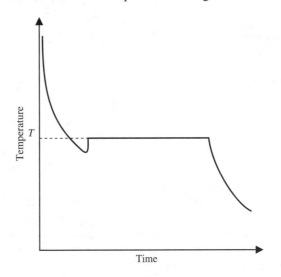

A cooling curve for a pure liquid shows a horizontal region at its melting temperature where the enthalpy of melting is being evolved.

When a cooling curve is plotted for a liquid mixture, it does not show a horizontal portion at a sharp freezing temperature as seen with a pure substance. Instead, there is a smaller change in the gradient of the curve, which occurs at the temperature at which solid begins to separate from the mixture. Typical cooling curves of two miscible liquids A and B, which do not form solid solutions in each other, are shown in Figure 2.9B.

A cooling curve for a liquid mixture shows a change in gradient at the temperature at which solid begins to separate.

The method of recording cooling curves for different liquid mixtures of metal A and metal B is as follows. Mixtures are prepared containing different percentages of A from 0% to 100%. One of the mixtures is melted in a nickel crucible. The source of heat is removed. A thermocouple in a protective sheath is inserted. The temperature is recorded every 15 seconds until the alloy is solid. The temperature is plotted against time [see Figure 2.9B]. The procedure is repeated for each alloy.

2.9.2 PLOTTING COOLING CURVES

A phase diagram is constructed by plotting this temperature against the compositions of the mixture.

From cooling curves [Figure 2.9B], the freezing temperatures (at atmospheric pressure) of mixtures can be obtained. A **phase diagram** can be constructed by plotting the freezing temperature at atmospheric pressure against the composition of the mixture [see Figure 2.9C].

In practice the lines from T_A to E and T_B to E are usually slightly curved because the two components A and B are slightly soluble in each other in the solid state [Figure 2.9C].

FIGURE 2.9B
Cooling Curves for Liquid
Mixtures of A and B

FIGURE 2.9C
A Phase Diagram for Two
Miscible Liquids A and B

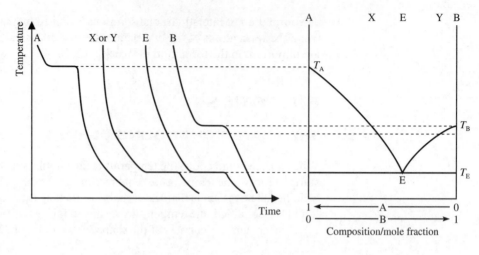

FIGURE 2.9D
Phase Diagram for Two
Metals which Form a
Eutectic Mixture

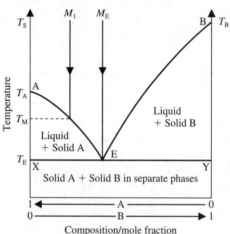

2.9.3 PHASE DIAGRAMS

A phase diagram for two metals which form a eutectic mixture is shown. The eutectic mixture is the mixture with the composition that has the lowest melting temperature of any mixture of the two components.

Figure 2.9D shows the phase diagram for mixtures of the metals A and B. Molten A and molten B are miscible in all proportions and do not form solid solutions in each other. T_A and T_B are the melting temperatures of pure A and pure B respectively. The curve AE shows how the melting temperature of pure A falls as increasing proportions of B are added to it. The curve BE shows how the melting temperature of B falls as A is added to it. E is the lowest temperature (at that pressure) at which mixtures of A and B can exist in the liquid state. It is called the **eutectic point** or **eutectic temperature**. The mixture with this composition (the proportions of A and B in the mixture that has the lowest melting temperature) is the **eutectic mixture**. The fixed composition and the sharp melting temperature make the eutectic mixture resemble a compound, but the fact that its composition is pressure-dependent shows that the eutectic is a mixture, not a compound. The heterogeneous structure can be seen under a microscope, when it appears as an intimate mixture of crystals of A and B. XY is a horizontal line through E. Below XY two solid phases, A and B are present. In the area AEX, solid A and a liquid mixture of A and B are present. In the area BEY, pure B is present with a liquid mixture of A and B. Above AEB, a single liquid phase exists; this is a mixture of A and B.

On cooling a mixture of the two components, the component which is present in excess of the eutectic composition crystallises from solution.

Consider what happens when a mixture M_1 is cooled. M_1 contains a greater proportion of A than the eutectic mixture. When liquid M_1 is cooled, it begins to deposit crystals of pure A at temperature T_M so that at this temperature solid A and liquid M_1 are in equilibrium. On further cooling more and more pure A is deposited, and the remaining liquid becomes poorer in A and therefore relatively richer in B. The changing composition of the liquid follows the curve AE. Eventually, the composition of the mixture reaches the eutectic mixture, E, when it all solidifies at the eutectic temperature. The curve AE represents the set of temperatures and compositions at which solid A and liquid M_1 are in equilibrium. A mixture which is richer in B than the eutectic mixture will deposit pure B on cooling until it reaches the eutectic composition, when all the liquid will solidify. Along the curve BE, solid B is in equilibrium with the molten mixture. The point E is the only point at which solid A, solid B and the liquid mixture are all three in equilibrium with each other. If a liquid eutectic mixture M_E is cooled, the mixture cools down to T_E and then solidifies completely. Observation of the cooling curve does not tell you whether M_E is a pure compound or a eutectic mixture. You can test by melting the solid, adding a little A or B and cooling again. If the temperature at which the first appearance of solid occurs is higher than before, the substance is a eutectic mixture.

On cooling a mixture of the eutectic composition, it solidifies completely at the eutectic temperature.

2.9.4 THE APPEARANCE OF SOLIDS WHICH SOLIDIFY FROM LIQUID MIXTURES

FIGURE 2.9E
The Appearance of Solids Obtained in Figure 2.9D

1 In the area AEX, solid A and a liquid mixture of A and B exist. Microscopic examination shows crystals of A in a solution of A and B.

2 In the area BEY, solid B and a liquid mixture of A and B exist. Microscopic examination shows crystals of B in a solution of A and B.

3 In the area below the line XY, solid A and solid B exist. Microscopic examination of a solid which forms by cooling a eutectic mixture shows a mixture of fine crystals of A and B.

4 When an alloy which is richer in A than the eutectic mixture is cooled, the solid consists of coarse crystals of A which form as the alloy cools down to T_E, and a mixture of fine crystals of A and B which forms when the eutectic mixture crystallises suddenly when T_E is reached.

5 When an alloy richer in B than the eutectic mixture is cooled, the solid consists of coarse crystals of B in a eutectic mixture of fine crystals of A and B.

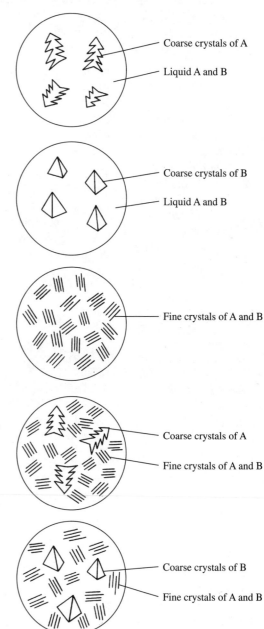

Coarse crystals of A

Liquid A and B

Coarse crystals of B

Liquid A and B

Fine crystals of A and B

Coarse crystals of A

Fine crystals of A and B

Coarse crystals of B

Fine crystals of A and B

2.9.5 HOW TO PREPARE METALS FOR MICROSCOPIC EXAMINATION

A metallurgical microscope differs from a biological microscope. Metals are opaque to light, and they must be illuminated by incident light – not transmitted light as in a biological microscope.

Specimens of metals must be prepared for microscopic examination. The surface is polished to remove scratches and cracks.

A metal specimen is obtained for microscopic examination by cutting a sample from a metal component. To prepare the metal for inspection, the surface must be rendered smooth, without cracks or scratches, and it must not be brightly reflecting. The first step in accomplishing this is to grind the surface of the metal with emery paper, working through a number of grades of emery paper from the coarsest to the finest grade. After all this polishing, the surface will be bright, although microscopic scratches will remain. These are removed by polishing with a commercial metal polish. The specimen is washed and dried and examined for scratches and cracks under a microscope. If necessary, polishing is continued. Then the metal is washed with soap solution and with propanone. The highly polished metal surface will appear bright and show no structure as it acts as a mirror to the incident light [see Figure 2.9F].

The surface must not reflect light brilliantly so it is dulled by etching with a suitable reagent . . .
. . . which attacks grain boundaries preferentially . . .
. . . and reveals details of the structure.

To reveal the structure of the specimen it is necessary to etch the polished surface. The choice of etching agent depends on the nature of the material. A solution of nitric acid in ethanol is often used for iron and steel, a solution of iron(III) chloride in hydrochloric acid is used for copper and its alloys, and aqueous sodium hydroxide for aluminium alloys. It is wise to proceed cautiously with this, etching for a few minutes, examining again and etching for longer if necessary. The etching agent will attack preferentially at crystal grain boundaries.

Finally the metal is washed with hot water and propanone. Examination of the etched specimen under the microscope will now reveal the grain boundaries. The etching agent does not generally dissolve the metal surface; it produces a faceted surface. The orientation of the surface facets may vary from grain to grain and the light reflected to the objective lens of the microscope will vary from one grain to another, making some grains appear lighter in colour than others [see Figure 2.9F].

FIGURE 2.9F
Microscopic Examination of a Metal Surface

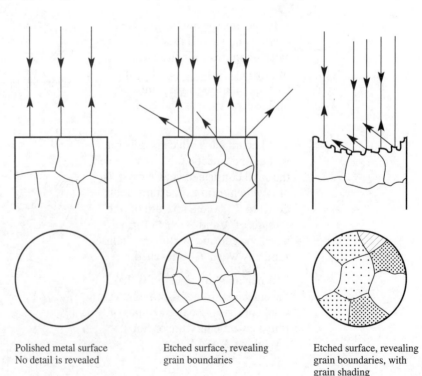

Polished metal surface
No detail is revealed

Etched surface, revealing grain boundaries

Etched surface, revealing grain boundaries, with grain shading

FIGURE 2.9G
Photomicrograph of
Alloy Showing Crystals in
a Eutectic Mixture

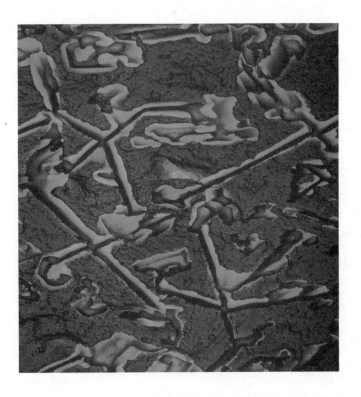

2.9.6 EXAMPLES OF EUTECTICS

Any two substances which are completely miscible as liquids but do not form a solid solution can form a eutectic mixture. Turbine blades are made from an alloy containing 80% nickel : 20% chromium. The eutectic mixture of iron and carbon (4.3% by mass carbon) is important because it determines the carbon content of the pig iron that comes from the blast furnace. The temperature–composition diagram for tin and lead shows how these metals form a eutectic mixture with a melting temperature lower than that of the individual metals. The low melting temperatures of mixtures of tin and lead account for the use of these alloys as solders. Table 2.9A lists the compositions of some eutectics.

The compositions of som eutectic mixtures are listed.

TABLE 2.9A
Examples of Eutectic
Mixtures

Elements	Melting temperatures/K			Composition of eutectic/ % by mass
	Element 1	Element 2	Eutectic	% (Element 1) : % (Element 2)
Aluminium–lead	933	599	519	12% : 88%
Zinc–copper	693	1356	90	30% : 70%
Aluminium–copper	933	1356	818	33% : 67%
Tin–lead	505	599	456	38% : 62%
Iron-carbon	1808	3925	1400	95.7% : 4.3%

CHECKPOINT 2.9

1. The compositions of some eutectic mixtures are tabulated.

Element A	Element B	Melting temperature/K			Mole fractions
		$T_m(A)$	$T_m(B)$	Eutectic	Element A : Element B
Cadmium	Bismuth	594	444	417	0.45 : 0.55
Silicon	Aluminium	1685	930	851	0.11 : 0.89
Beryllium	Silicon	1555	16 850	1360	0.68 : 0.32

(*a*) A mixture of cadmium and bismuth in which the mole fraction of bismuth is 0.65 is heated to 600 K and allowed to cool slowly. Describe the composition of the solid material in the crucible at room temperature.

(*b*) Sketch the temperature–composition graphs for silicon–aluminium mixtures.

(*c*) A mixture of 1.00 mol aluminium and 1.00 mol silicon is melted and heated to 1700 K and allowed to cool slowly. State the names and the amounts of the solids present in the crucible at room temperature.

(*d*) Sketch the phase diagram for beryllium–silicon mixtures. Say what happens when the following liquids are cooled from 1700 K: (i) 9.0 g silicon in 6.1 g beryllium, (ii) 10.0 g silicon in 6.1 g beryllium, (iii) 6.0 g silicon in 6.1 g beryllium.

2.

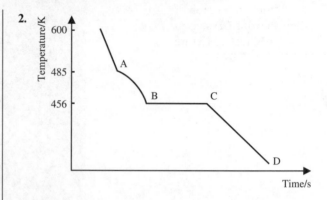

The curve shown in the figure is a cooling curve for an alloy of tin and lead with the mole fractions 0.90 tin : 0.10 lead. Melting temperatures are T_m (tin) = 505 K, T_m (lead) = 600 K.

(*a*) Explain what is happening at the parts of the curve labelled (i) AB and (ii) BC.

(*b*) Describe how you would determine this cooling curve in the laboratory.

(*c*) Sketch the microstructure of the solid formed at D.

2.10 MECHANICAL TREATMENT AND HEAT TREATMENT

2.10.1 WORK HARDENING

Metals can be cold worked with an increase in strength and a decrease in ductility ...

... a process called work hardening ...

... in which the metals become more rigid, more difficult to reshape.

Metals can be 'cold worked': forced into new shapes at a temperature below the recrystallisation temperature [see definition in § 2.10.3]. Eventually cold-working makes metals work-hardened and difficult to reshape. The forces which are applied to work the metal produce additional dislocations. At first the dislocations move easily as the shape of the metal changes, but as working continues, the movement of dislocations becomes difficult [see § 2.7]. The metal eventually reaches a state where strength and hardness are at a maximum; beyond this point the metal becomes too brittle [see § 1.4]. Sometimes the rigidity of work-hardened metal is advantageous. The front wheel arch of a motor vehicle is much more rigid than the sheet of metal from which it was formed. Cold working increases strength and also reduces ductility. Cold working below 0.3 T_m (T_m = melting temperature in kelvins) is used for small accurate changes in shape [see § 2.11].

2.10.2 RECOVERY

When a work-hardened metal is heated, atoms diffuse and rearrange within grains.

In a work-hardened metal, the atoms are locked in a state of high potential energy because of the distortions in the crystal structure. (Potential energy is the energy which an object has due to its position, in this case the positions of the atoms relative to one another.) If the work-hardened metal is heated, the atoms may gain enough energy to

This treatment is called recovery or stress relief.

diffuse and to rearrange within the grains. The period of heating must be short and the temperature must not be high enough to allow the grains to grow. This heat treatment is called **recovery** or **stress relief**.

2.10.3 RECRYSTALLISATION

If the temperature is raised up to the recrystallisation temperature new grains will grow, restoring ductility.

If the temperature is raised above that needed for recovery, new grains will grow. Work-hardened grains have high potential energy, and at a high enough temperature will recrystallise as numerous fine grains. The temperature at which new grains begin to grow is called the **recrystallisation temperature**. In general, the recrystallisation temperature is approximately equal to $0.5\,T_{\mathrm{m}}$. Recrystallisation leads to a marked change in properties, an increase in ductility and a reduction in strength.

2.10.4 ANNEALING

Annealing restores work-hardened metals to a ductile condition. The metal is put in a furnace until recovery and recrystallisation are complete.

Work-hardened metals can be restored by a treatment called **annealing**. The metal object is put into a furnace and left until both recovery and recrystallisation are complete. The new recrystallised grains make the annealed metal softer and more malleable and ductile because the dislocations are mobile. When a metal is shaped by cold working, it may need to be annealed several times during the process. Each time work-hardening makes the metal too hard for further cold working, it is put into the annealing furnace until recovery and recrystallisation are complete.

2.10.5 HOT WORKING

Hot working or forging is working a metal above its recrystallisation temperature. Annealing takes place while the metal is worked.

In **hot working** or **forging** a metal is worked above its recrystallisation temperature. In this way annealing is taking place while the metal is worked. Forging requires less mechanical energy than cold working because the metal is softer and because there are no interruptions for annealing. Obviously, more heat energy is required for hot working than for cold working. Hot working above $0.6\,T_{\mathrm{m}}$ (rolling and forging) is used for major changes in shape [see §2.12].

2.10.6 QUENCHING AND TEMPERING

Quenching – the sudden cooling of a hot metal to room temperature – ...

... produces a large number of fine precipitates ...

... which increase the strength of the metal ...

... by making the grains harder ...

... and by restricting the movement of dislocations.

Quenching is the sudden immersion of a hot metal in oil or water at room temperature. It is an important treatment for cutting tools. Quenching is most used for alloys in which one metal is dissolved in another to form a solid solution. At high temperatures, more of the solute can be held in solution. If an alloy is heated and cooled slowly, the solute precipitates out slowly. If it is quenched, there is not time for precipitation to occur, and a supersaturated solution is formed. The excess of solute metal is trapped within the grains, making them harder. The type of heat treatment controls the amount and particle size of precipitates. These precipitates influence the movement of dislocations. A large amount of fine precipitate gives a bigger increase in strength than a smaller amount of coarser precipitate.

Many cutting tools are made of carbon steel containing slightly less than 1% carbon. At 750 °C pure iron exists in the **austenite** form, a body-centred-cubic structure which is not close-packed [see §2.5]. The crystal structure contains sites between the iron

When steel is cooled, the crystal structure of iron changes from the austenite structure to the ferrite structure.
The austenite structure can hold more carbon than the ferrite structure.
Quenching 'freezes' the austenite structure, trapping carbon . . .
. . . making quenched steel hard and brittle.

atoms that will accommodate carbon atoms and hold 1% of carbon in solid solution. As the steel cools, it changes from the austenite form into the **ferrite** form, a close-packed face-centred-cubic structure in which the sites between iron atoms are smaller and can hold fewer carbon atoms. On slow cooling, carbon atoms have time to vacate the sites, and carbon leaves the austenite crystal structure, with the formation of the alloy **cementite**. On rapid cooling the carbon atoms are trapped in the crystal structure and an alloy called **martensite** is formed. The distortion of the structure by the presence of carbon atoms impedes the movement of dislocations. It becomes very difficult to deform the metal, and the metal is hard and brittle. When steel is quenched rapidly, the austenite form, with its higher carbon content, persists at the lower temperature. This is why quenching is employed for hard steels which are to be used for cutting tools.

Quenched steel is hard but it is too brittle, and to make it less brittle, some hardness must be sacrificed. To obtain a balance between malleability and hardness, the steel is tempered. The metal object is heated a second time to a temperature between 200 and 300 °C. The steel shows its 'temper colours', ranging from pale straw colour at 200 °C to dark blue at 300 °C, owing to an increasing film of oxide. After reaching the required **tempering temperature** the steel is quenched again. The temperature at which the steel is worked must not be above the tempering temperature. Tempered steel is hard and malleable.

Tempered steel is hard and malleable . . .
. . . made by heating quenched steel to the tempering temperature and quenching again.

Metals and ceramics undergo transformations in crystalline structure. Polymers rarely undergo such solid state transformations except by forming thin, close-packed crystalline regions through chain alignment or chain folding [see § 4.3].

CHECKPOINT 2.10

1. Explain the terms recovery, recrystallisation, annealing, quenching. Describe the changes that take place in the structure of a metal during each of these processes.

2. What is tempering? What effects does it have on (*a*) the structure and (*b*) the physical properties of a metal?

3. Austenite has a body-centred-cubic crystal structure, while ferrite has a face-centred-cubic structure. How does this explain why the austenite form of iron is able to hold more carbon in its crystal structure than the ferrite form? What happens to the carbon that is released from the crystal structure when austenite is cooled rapidly?

4. (*a*) Give two advantages of the hot working of metals over cold working.

    (*b*) Give two advantages of cold working over hot working.

2.11 METHODS OF SHAPING METALS

2.11.1 CASTING

Metals can be shaped by casting – pouring liquid metal into a mould and allowing it to cool and solidify.

Most metal objects have been cast at some stage of manufacture. Casting is shaping an object by pouring liquid metal into a mould and allowing it to cool and solidify. The shape may be that of the final object or it may need finishing by machining. As the metal cools, it contracts a little and the object will be slightly smaller than the inside of the mould. The mould must be designed so that the liquid metal flows into all parts of the mould. It must also allow the object to be released, so moulds are made in two or more parts which can be held together when the metal is poured in and separated when the metal has solidified.

Sand casting is a method in which the mould is made of sand and clay. The mould is made by packing a mixture of sand and clay round a pattern of the required object. The mould is made slightly larger than the object and in two parts.

FIGURE 2.11A
Sand Casting

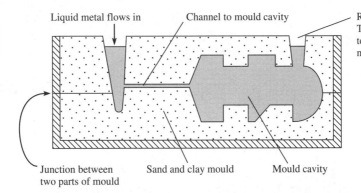

Liquid metal flows in   Channel to mould cavity   Riser holds liquid metal. This liquid metal flows down to fill the gap as the moulded metal contracts.

Junction between two parts of mould   Sand and clay mould   Mould cavity

Die casting involves the use of a metal mould. **Gravity die casting** is similar to sand casting in that the liquid metal is poured into the mould. **Pressure die casting** injects liquid metal into the mould under pressure. It is better at forcing metal into all parts of the mould and is the method chosen for complicated shapes. **Centrifugal casting** is a technique which is used for simple shapes [see Figure 2.11B].

FIGURE 2.11B
Centrifugal Casting

*Methods include . . .
. . . sand-casting . . .
. . . die-casting . . .
. . . gravity die-casting . . .
. . . pressure die-casting . . .
. . . centrifugal
die-casting . . .
. . . investment casting or
lost wax casting.*

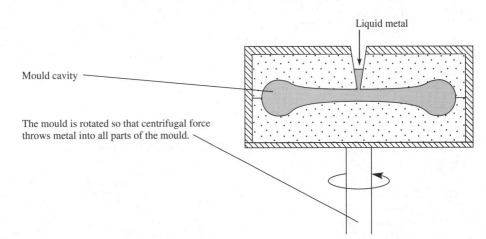

Liquid metal

Mould cavity

The mould is rotated so that centrifugal force throws metal into all parts of the mould.

Investment casting or **lost wax casting** is used for metals of high melting temperature. First, a metal mould is used to make a wax pattern. Then the wax pattern is coated with a ceramic paste and heated to harden the ceramic and melt the wax. A ceramic mould has been formed, and liquid metal is injected into this under pressure. When the metal has cooled and solidified, the ceramic is broken to release the casting. The method is the only one suitable for high-melting metals which soon ruin metal dies.

2.11.2 GRAIN STRUCTURE

The different casting methods give different grain structures. When a molten metal is poured into a mould, the metal in contact with the mould cools faster than that in the centre. Small **chill crystals** form. As the metal in the centre cools more slowly, some crystals grow inwards as long **columnar crystals** perpendicular to the walls of the mould. In the centre of the melt, the liquid metal is in motion owing to convection currents, and the crystals which form are medium-sized, almost spherical crystals called **equiaxed crystals**.

*Different casting methods give different grain structures.
On the outside of the cast, small chill crystals form.
Some extend inwards as columnar crystals.
In the centre of the melt, small spherical equiaxed crystals form.*

Castings with small equiaxed crystals are preferred. This type of structure is promoted by rapid cooling. Sand castings cool slowly and tend to have large columnar crystals and therefore relatively low strength. Die casting with metal moulds gives faster cooling and a larger zone of equiaxed crystals, making for high strength. The inclusion of additives promotes the growth of finer grains.

FIGURE 2.11C
A Section of a Casting

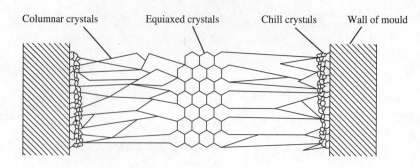

Columnar crystals   Equiaxed crystals   Chill crystals   Wall of mould

*A casting that contains
equiaxed crystals is
preferred.*

2.11.3 METHODS OF COLD WORKING

Cold working (that is, below the recrystallisation temperature) may be done by methods which include rolling, drawing, deep drawing, pressing [see Figure 2.11D] and extrusion (as for hot extrusion, Figure 2.11F).

FIGURE 2.11D
(a) Rolling, (b) Drawing,
(c) Deep Drawing,
(d) Pressing

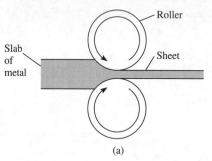

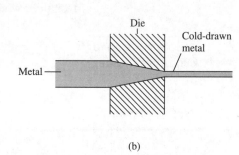

(a)

(b)

Rolling: A slab of metal is rolled into a sheet.

Drawing: A metal is pulled through a die (a pattern).

*Cold working can be
done by:
rolling,
drawing,
deep drawing,
pressing,
extrusion.*

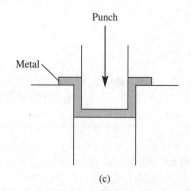

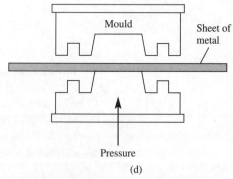

(c)

(d)

Deep drawing: A punch forces a blank to flow into a die in the forming of tubes, cans, etc.

Pressing: Making a car body panel by pressing a sheet of metal on to a die

2.11.4 METHODS USED IN HOT WORKING

Hot working (that is, above the recrystallisation temperature) uses methods which include the following.

1. Rolling at a temperature of $0.6\,T_m$ where T_m is the melting temperature in kelvins

2. Forging, either hammering a ductile metal or squeezing it between a pair of dies [see Figure 2.11E]

3. Extrusion at a temperature between $0.65\,T_m$ and $0.9\,T_m$ [see Figure 2.11F]. Cold extrusion gives a work-hardened metal; hot extrusion gives a soft ductile metal. If the diameter of the metal is to be greatly reduced, hot extrusion is required.

FIGURE 2.11E
Forging

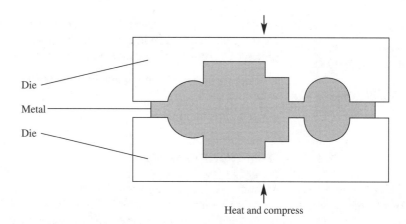

Die

Metal

Die

Heat and compress

FIGURE 2.11F
Extrusion

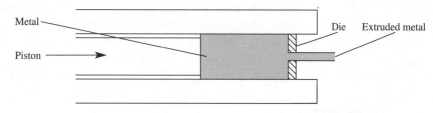

Metal

Die   Extruded metal

Piston

*Hot working may be
done by:
rolling,
forging,
extrusion,
sintering.*

4. Sintering, in which a metal component is made from a metal powder. The metal powder is compacted in a mould and heated to a temperature at which the particles combine. The process is useful for brittle metals and for high melting temperature metals for which casting is expensive and also for ceramics [see § 3.2].

CHECKPOINT 2.11

1. Methods which are used for casting metal objects are sand casting and die casting. How do the two methods compare?

(*a*) Which gives the stronger metal object? Why?

(*b*) Which is better for castings of complicated shapes?

(*c*) What techniques are used in each case to ensure that the molten metal fills the mould?

(*d*) Which method is better for high melting temperature castings, e.g. steel ingots?

(*e*) Which method is used for low melting temperature castings, e.g. cast iron?

2. Suggest a method that could be used to form metal into each of the following.

(*a*) a sheet of foil, e.g. aluminium foil

(*b*) a length of wire, e.g. copper wire

(*c*) a mild steel car door

(*d*) a tin-plated steel can to hold a soft drink

(*e*) an iron car engine block.

2.12 CORROSION

*Corrosion is the change of
a metallic element into a
compound.*

The causes of metal failure, including fatigue failure, were discussed in § 1.12. One of the causes of failure is **corrosion**. Corrosion is a reaction in which a metal changes from the element into a compound. For the majority of metals, the combined state is the stable state. For most metals, the standard free energy of formation of the compounds is negative [see *ALC*, § 10.9.1]. The metals are therefore found naturally occurring in the combined state, and corrosion is a natural process. Metals and alloys react with substances in the environment to form compounds. The product of corrosion is often an oxide since metals can react with oxygen in the air. Other substances in the atmosphere which cause corrosion are sulphur dioxide, carbon dioxide and chlorine. In industrial environments other compounds also may be present. Corrosion may be divided into two classes:

- dry corrosion in which a metal or alloy reacts with a gas or gases in the absence of water
- wet corrosion in which a metal or alloy reacts with an aqueous environment

*For most metals, the combined state is the stable state . . .
. . . and corrosion is a natural process.*

2.12.1 DRY CORROSION

In dry corrosion metals are oxidised by the oxygen of the air.

Most metals tend to be oxidised at room temperature and a freshly cut metal surface therefore soon becomes coated with a film of oxide. The film separates the metal from the atmosphere so that further oxidation can occur only if the film breaks down or if oxygen or metal is able to diffuse through it. The metals in Groups 1 and 2 form oxide films which tend to be porous and offer poor protection. In the metals which are chosen for use in construction and engineering, e.g. iron and aluminium, the oxide film continues to grow, following a parabolic curve [see Figure 2.12A]. For this to occur there must be diffusion of the metal or oxygen or both through the oxide film.

FIGURE 2.12A
The Growth of an Oxide Film on a Metal

Oxidation can proceed only if the initial oxide film formed on the surface of the metal is permeable to

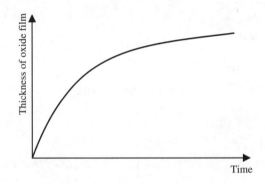

If the oxide were of fixed stoichiometric composition, diffusion could not occur, and oxidation would cease. For dry corrosion to occur, the oxide must contain certain **defects**, which make it non-stoichiometric. The film which forms on the surface of iron has formula $Fe_{(1-x)}O$. Since it contains fewer Fe^{2+} ions than O^{2-} ions, there must be **structural defects** – vacant sites in the crystal structure where some Fe^{2+} ions are missing. The oxide must nevertheless be electrically uncharged and this is achieved by the presence of some Fe^{3+} ions in the lattice. These ions in a higher oxidation state (Fe^{3+} ions) are described as **electronic defects** in the iron lattice [see Figure 2.12B].

The oxides of metals in Groups 1 and 2 are porous and do not protect the metals from further oxidation.

FIGURE 2.12B
Structure of $Fe_{(1-x)}O$.

---

*In e.g. iron and aluminium, the oxide film continues to grow and form a protective layer. Oxygen is able to diffuse through the oxide layer because the oxide layer contains defects: structural defects, e.g. missing Fe^{2+} ions . . .
. . . electronic defects, e.g. Fe^{3+} ions.*

Fe^{2+}	O^{2-}	Fe^{2+}	O^{2-}	Fe^{2+}	O^{2-}	Fe^{2+}	O^{2-}
O^{2-}	Fe^{2+}	O^{2-}	Fe^{3+}	O^{2-}	Fe^{2+}	O^{2-}	Fe^{2+}
Fe^{3+}	O^{2-}		O^{2-}	Fe^{3+}	O^{2-}	Fe^{2+}	O^{2-}
O^{2-}	Fe^{2+}	O^{2-}	Fe^{3+}	O^{2-}	Fe^{2+}	O^{2-}	Fe^{2+}

A structural defect: a vacant site in the structure

An electronic defect

Iron(II) ions can diffuse from the metal through the vacant sites in the oxide layer to meet oxygen at the surface. Electrons can diffuse through the oxide layer by switching from Fe^{2+} to Fe^{3+} and the reverse. At the surface, electrons meet oxygen and convert it into oxide ions. In this way, more iron oxide is formed at the boundary between metal oxide and oxygen. The flow of electrons is more likely in elements such as iron which show more than one oxidation state. Resistance to corrosion can be improved by alloying with an element which remedies the defects in its structure.

2.12.2 WET CORROSION

In wet corrosion, the metal releases electrons to form ions ...
... the electrons combine with hydrogen ions to form hydrogen molecules.

Consider the corrosion that occurs when a metal M is attacked by an acid.

$$M(s) \longrightarrow M^{n+}(aq) + ne^-$$

$$2H^+(aq) + 2e^- \longrightarrow H_2(g)$$

The areas in which **oxidation** occurs are **anodic areas**: the metal releases delocalised electrons from its metallic bonds, and these electrons flow from anodic areas to cathodic areas. At **cathodic areas, reduction** occurs, that is, hydrogen ions accept electrons to form hydrogen molecules. The process continues as long as there are hydrogen ions to accept the electrons released by the metal. Consider the corrosion of iron in a neutral salt solution. Iron is oxidised:

$$Fe(s) \longrightarrow Fe^{2+}(aq) + 2e^-$$

If the reaction is to continue, electrons must be removed; therefore in the absence of oxygen and the absence of hydrogen ions, the reaction soon comes to a halt. In the presence of oxygen, oxygen molecules accept electrons to form hydroxide ions:

$$O_2 + 2H_2O + 4e^- \longrightarrow 4OH^-$$

Iron(II) hydroxide, $Fe(OH)_2$, is then precipitated. In air, iron(II) hydroxide is oxidised to a hydrated iron(III) oxide, rust.

Corrosion is accelerated by the presence of a salt solution which acts as an electrolyte ...
... and by the association of two metals to form a galvanic cell.

Rusting takes place more rapidly in sodium chloride solution than in water. This is because the salt improves the conductivity of the electrolyte and assists the conduction of ions through the electrolyte. The coupling of two metals of different standard electrode potential in a solution leads to corrosion. The further apart the metals are in the electrochemical series, the more rapidly does corrosion occur. [For the electrochemical series, see *ALC*, § 12.3.1.] What happens when iron is coupled with a metal lower in the electrochemical series? Imagine that brass screws (70% copper and 30% zinc) have been used in an iron structure. Since copper is lower in the electrochemical series than iron, copper can act as the cathode in a galvanic cell with iron as the anode.

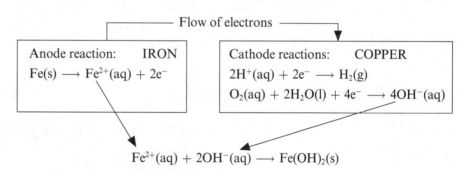

The Fe^{2+} ions formed at the anode and the OH^- ions formed at the cathode meet in between to form a precipitate of iron(II) hydroxide. This reacts with air to form rust, hydrated iron(III) oxide. Corrosion continues until one of the metals is used up.

Factors which increase the rate of corrosion include:
... a corrosive environment ...
... a surface defect ...
... stress ...
... a rise in temperature ...
... the formation of a galvanic cell with a second metal.

Factors which influence the rate of corrosion are:

1. Environment: Water and air must be present. A marine environment is particularly corrosive. Chemical pollutants such as sulphur dioxide increase the rate of corrosion.

2. Surface defects: If a protective coating has been damaged, water may come into contact with the metal.

3. Stress: If a component has a heavily cold-worked part and the rest of it less well worked, the heavily cold-worked part will act as anode and the less-worked part as cathode.

4. Temperature: A rise in temperature increases the rate of corrosion.

5. Other metals: A galvanic cell can be formed by a combination of two metals.

2.13 CORROSION PREVENTION

Methods of preventing corrosion never work perfectly, but they can retard the progress of corrosion. The methods widely used to protect iron and steel against corrosion are:

Methods of preventing corrosion of iron and steel include . . .
. . . a coating of paint, oil or grease . . .

1. Apply a coat of paint. This is very effective for large structures, e.g. bridges, but not for moving components. If the paint is scratched, the iron underneath rusts. Enamel is also used.

2. Apply a coating of oil or grease. This method is used for moving parts of machinery.

3. Use a coating of a metal which is lower than iron in the electrochemical series, e.g. tin, as in food cans. If the coating is scratched, the iron underneath rusts.

4. Use a coating of a metal which is higher than iron in the electrochemical series, e.g. zinc, as in galvanised iron. This coating is a **sacrificial coating**, which will eventually be corroded while the iron beneath is protected.

. . . a coating of a metal lower in the electrochemical series . . .
. . . a coating of a metal higher in the electrochemical series which is a sacrificial coating . . .

5. Use a sacrificial anode. The sacrificial metal need not be in the form of a coating; it may be a **sacrificial anode**. As long ago as 1842 Sir Humphrey Davy had the idea of attaching large pieces of zinc to ships' hulls. The pieces of zinc act as anodes in galvanic cells and are corroded, while the steel of the hull acts as the cathode. This method is described as **cathodic protection**. The blocks of zinc, which corrode in preference to iron, can be easily replaced, while the valuable structure, the ship, is preserved.

6. Apply an external voltage to reverse the normal flow of electrons so that iron becomes the cathode. This method of cathodic protection is expensive and is not used for large objects.

7. Use an oxide coating. Stainless steels are alloys of iron with chromium or chromium and nickel. The chromium in the steel reacts with air to form a layer of chromium oxide which protects the steel from corrosion.

8. Apply rust inhibitors. A rust inhibitor used in a car radiator contains chromates and phosphates which are adsorbed on the metal surface and insulate it from the electrolyte. For rusting and prevention, see *ALC*, §§ 24.14.4 and 24.14.5.

FIGURE 2.13A
Sacrificial Anodes on the Hull of a Ship

. . . attaching a sacrificial anode . . .
. . . applying an external voltage . . .
. . . using a stainless steel . . .
. . . applying a rust inhibitor.

================================ CHECKPOINT 2.13 ================================

1. Cars in dry desert parts of the USA remain free of rust compared with cars in a damp climate. Suggest an explanation.

2. The water in a domestic boiler is de-aerated. Why does this reduce corrosion?

3. Two metals must be joined in an engineering project. In order to keep corrosion to a minimum, what criterion must be observed in the choice of metals?

4. Aluminium pipes are used to carry water into a tank. The tank may be made of galvanised steel or of copper. Which of these would keep corrosion to a minimum?

5. Pieces of magnesium are placed close to buried iron pipes. Explain why this reduces corrosion of the iron.

6. In a junction between mild steel and copper in a marine environment, the mild steel corrodes in preference to the copper. In a mild steel–aluminium junction in the same environment, the aluminium corrodes more than the mild steel. Explain these observations.

7. Suggest methods that could be used to protect a steel water storage tank.

8. Explain how it is possible for a metal which forms an oxide film when a fresh surface is exposed to the air to corrode further.

2.14 RECYCLING METALS

Recycling metals conserves metal ores and saves energy.

In order to produce iron, iron ore must be mined, transported and smelted. This takes several times more energy than collecting and remelting scrap iron. Aluminium is a prime candidate for conservation for two reasons. Owing to its very high resistance to corrosion, used aluminium is as good as new. A second reason is the high energy consumption in the extraction of aluminium from bauxite [see *ALC*, § 19.2.2]. The cost of reusing scrap aluminium is only one twentieth of the cost of making the pure metal.

The collection, sorting and recycling of metals has become an important industry. A major source of used metals is motor vehicles. There are about 140 million cars in Europe and 12 million a year reach the end of the road. About 75% of this mass of about 12 million tonnes is iron and steel. The European Community has encouraged member nations to make car recycling a priority. What is needed is to be able to identify components and to be able to dismantle cars rapidly. Some German car manufacturers, e.g. BMW and Mercedes, have started to design recyclability into their vehicles. They have developed **disassembly lines**.

The first step in disassembly is to remove petrol, oil and brake fluid. Then heavy items such as batteries, cylinder blocks, gearboxes, etc. are sent to specialist firms. A fragmenter smashes the rest of the material into small pieces. A magnet separates most of the ferrous metal. The rest passes to a second separator [see Figure 2.14A].

FIGURE 2.14A The Eriez Cotswold Separator

Some firms have started to dismantle old cars . . . . . . separate the metals in them . . . . . . and recycle the metals.

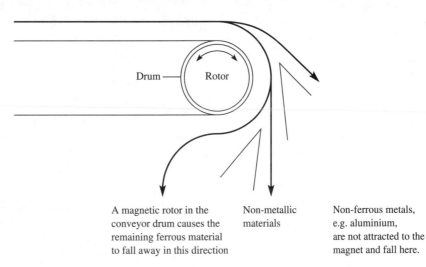

Drum — Rotor

A magnetic rotor in the conveyor drum causes the remaining ferrous material to fall away in this direction

Non-metallic materials

Non-ferrous metals, e.g. aluminium, are not attracted to the magnet and fall here.

The non-magnetic material is a mixture of rubber, glass, plastic and non-ferrous metals. It is dealt with by a **sink-and-float process**. The material is flooded with water which carries away dirt and textiles. The scrap then passes to tanks containing solutions of density ranging from 1.25 to 3.8 kg dm^{-3}. Upholstery floats while metals sink.

The non-ferrous metals also are separated by the sink-and-float process. In a tank containing a solution of high density, aluminium floats and other metals sink to be recovered later. The reclaimed aluminium is pure enough to be used for making the alloys which the automobile industry uses.

QUESTIONS ON CHAPTER 2

1. Explain how the electrical conductivity of metals results from their electronic structure and bonding.

2. Explain how the ductility of metals is related to

(*a*) the nature of the metallic bond

(*b*) the presence of dislocations

(*c*) grain size

(*d*) the presence of 'foreign' atoms.

3. Explain in terms of dislocations: (*a*) work hardening, (*b*) recovery, (*c*) annealing.

4. What tests could be done to find out

(*a*) which of two specimens of steel is surface-hardened and which is not

(*b*) which of two specimens of steel has been tempered and which has not

(*c*) which of two specimens of steel is easier to form by bending. (Refer to Chapter 1 if necessary.)

5. Alloys have physical properties which differ from those of the constituent metals. Describe these differences and explain why alloying has these effects.

6. The strength of a material may be measured by the yield stress or the tensile strength or the fracture stress. Say where each of the terms is most useful.

7.

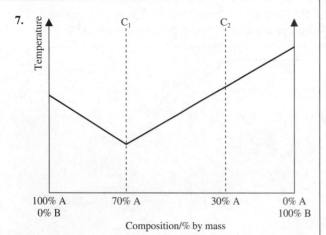

| 100% A | 70% A | 30% A | 0% A |
| 0% B | | | 100% B |

Composition/% by mass

The graph shows how the melting temperature of a mixture of two metals A and B depends on the composition. A and B do not form solid solutions or compounds with each other.

(*a*) Explain what happens on cooling (i) a mixture of composition C_1, (ii) a mixture of composition C_2.

(*b*) The solids which form when C_1 and C_2 are cooled were etched and examined under a microscope. Describe the appearance of the two solid surfaces.

8. (*a*) Explain what is meant by the statement that dry corrosion of a metal depends on defects in its crystal structure.

(*b*) Explain why the use of a *sacrificial* anode is described as *cathodic* protection.

(*c*) Suggest an alternative method of cathodic protection.

9. The allotrope of pure iron which is stable between 1390 and 910 °C is austenite. The allotrope which is stable below 910 °C is ferrite. One of these forms has a face-centred-cubic structure and the other has a body-centred-cubic structure.

(*a*) Which of these two structures is (i) the more closely packed, (ii) the structure of austenite?

(*b*) Sketch (i) the face-centred-cubic structure and (ii) the body-centred-cubic structure. State the coordination number in each.

(*c*) How does the transition from austenite to ferrite affect the ability of iron to combine with carbon?

10. (*a*) Distinguish between interstitial solid solutions and substitutional solid solutions.

(*b*) Distinguish between elastic and plastic deformation of metals.

(*c*) Sketch the change in the crystal structure of a metal that accompanies (i) elastic strain, (ii) permanent deformation.

11. Car exhaust systems need replacing many times during the lifetime of a car.

(*a*) What conditions are exhaust systems subject to?

(*b*) What properties do the materials used for exhaust systems require?

(*c*) Suggest suitable materials.

(*d*) What materials are generally used?

12. Which of the following metals are good candidates for recycling? Explain your answer.

(*a*) aluminium used in food containers

(*b*) tungsten used in the tips of drills

(*c*) lead in car batteries

(*d*) zinc used in protecting offshore oil rigs from corrosion

(*e*) tungsten filaments used in electric light bulbs

(*f*) iron in rusted car bodies.

3

CERAMICS

3.1 WHAT ARE THEY?

The blast furnace in which iron ore is smelted is lined with refractory (heat-resisting) bricks. Without crucibles and furnaces of such refractory materials, which can contain metals at their melting temperatures, it would have been impossible to develop the metallurgical operations that give us metals and alloys. These refractory materials are **ceramics** and have been known and used for thousands of years.

Space-age ceramics

When a space shuttle re-enters Earth's atmosphere from space, the friction that it generates is converted into heat and the temperature of the surface can reach 1500 °C – hotter than molten steel. A material must be found to insulate the interior of the space shuttle so as to protect the computers and recording instruments inside. A covering layer of heat-resistant tiles made of ceramics does the job. A tile consists of very fine coated silica fibres packed into an open cellular structure. The fibres are so loosely packed that 95% of the volume is air. The tiles have about the same density as cotton wool. Cellular materials are poor conductors of heat, and these ceramic tiles are probably the most heat-resistant materials ever made.

FIGURE 3.1A
A Space Shuttle

---

*Ceramics have been in use for thousands of years . . .
. . . and now have new space-age uses.
The ceramic tiles which coat space shuttles are the most heat-resistant material ever made.*

Ceramics are an important class of materials. Brick and pottery are ceramics. Traditional ceramics such as these are derived from the raw materials clay and silica and are essentially **silicates**. They are formed on, for example, a potter's wheel and hardened in a fire kiln. The name comes from the Greek *keramos*, meaning potter's earth.

Traditional ceramics include brick and pottery.

In recent years new ceramics have been developed to meet the demand for materials which withstand high temperatures and high pressures and resist chemical corrosion. Many ceramics have good mechanical strength, and are hard and resistant to wear. Some have special electrical characteristics. Their low conductivity of heat and electricity makes them useful as building bricks and in electrical insulation. Ceramics are important components of the magnetic materials used in communications. Ceramic magnets, e.g. Magnadur, are used for the ring magnets in television receivers. Ceramics have engineering and electronic applications. Examples of ceramics are:

- Whitewares: china, earthenware, pottery, porcelain, stoneware, vitreous ware
- Structural clay products: building brick, face brick, terracotta, sewer pipe, drain tile
- Refractory ceramics: firebricks, silica, silicon carbide, zirconium oxide, aluminium silicate
- Specialised ceramics
- Enamels and enamelled metals
- Glasses

Modern ceramics are valued for ... ... being able to withstand high temperature and high pressure ... ... resistance to chemical corrosion ... ... good mechanical strength ... hardness ... resistance to wear ... ... thermal and electrical insulation ... ... and in some cases magnetic properties. They have many important uses.

Ceramics in the diesel engine

Modern ceramics, e.g. silicon nitride and silicon carbide, can function at higher temperatures than alloy steels and are used as components of the internal combustion engine. The inclusion of cermet (ceramic and metal) parts and ceramic parts and insulating ceramic coatings can raise the operating temperature of a diesel engine from 700 to 1100 °C, which increases the engine's efficiency by 50%. These ceramics are made by compressing hot mixtures of finely powdered oxides. Alternatively the powders may be sprayed on to a metal to give a hard coating.

3.1.1. A DEFINITION OF CERAMICS

A **definition of ceramics** is the following. Ceramics comprise all inorganic engineering materials and products, except for metals and alloys. Ceramics are usually made through high temperature processing. The word **ceramic** can be used as a noun and also as an adjective meaning inorganic and non-metallic. There are composite materials which combine metals and ceramics and others which combine ceramics and organic materials; these are called **ceramic materials**.

Many ceramics are crystalline materials: they consist of an ordered arrangement of atoms. Many are compounds of metallic and non-metallic elements, e.g. silicon or aluminium combined with oxygen, carbon or nitrogen. On the basis of structure, ceramics can be classified into four main groups:

Ceramics include all inorganic engineering materials ... ... except for metals and alloys. They are crystalline compounds of metals and non-metallic elements.

CRYSTALLINE CERAMICS

These may be single substances such as magnesium oxide, aluminium oxide, zirconium oxide, titanium carbide, zirconium carbide, tungsten carbide, silicon carbide and silicon nitride. They may be mixtures of materials such as these. In magnesium oxide the bonding is ionic; in silicon carbide the bonding is covalent; in aluminium oxide the bonds have partial ionic character and partial covalent character.

AMORPHOUS CERAMICS

These are glasses, in which the arrangement of atoms has a degree of disorder. Examples are soda glass, silica glass and Pyrex glass.

BONDED CERAMICS

These are materials in which individual crystals are bonded together in a glassy matrix. The products derived from clay fall into this category, e.g. building bricks, terracotta, pottery and china.

CEMENTS

Some of these are crystalline, e.g. calcium silicate, and others are mixtures of crystalline and amorphous phases, e.g. concrete.

3.1.2 STRENGTH

Ceramics are brittle, with low tensile strength and low toughness. They can fail at low stress. Because the toughness is so low, small defects, e.g. surface scratches or lumps, can reduce the strength drastically. On the other hand, ceramic fibres, e.g. glass fibres, aluminium oxide fibres, silicon carbide fibres, are among the strongest materials known. This is because their surface defects are less than 1 μm in length. The effect of a defect in structure on the strength of a brittle material was covered in the discussion of Griffiths cracks in § 1.10. For the strength of a ceramic to be sufficient for use as an engineering ceramic the defect length must be very small – a few micrometres only.

Ceramics are brittle ...
... with low tensile strength and low toughness.

Research work continues on the problems of low tensile strength and low toughness of ceramics. Solutions to the problems lie in composite materials [see § 1.2 and Chapter 5]. The search is on for fibres which will reinforce ceramics. The hope is to achieve the same sort of success as fibre-reinforced plastics [see § 5.3].

3.2 MANUFACTURE

The three main raw materials used in the manufacture of ceramics are clay, sand and feldspar (aluminium potassium silicate, $K_2O \cdot Al_2O_3 \cdot 6SiO_2$). Clays are impure hydrated aluminium silicates. They have been formed as the result of weathering of igneous rocks which contained feldspar. An example is kaolinite, $Al_2O_3 \cdot 2SiO_2 \cdot 2H_2O$. Other clays contain potassium, magnesium and calcium. Clays become plastic and mouldable when they are finely pulverised and wet. They become rigid when dry and vitreous (glassy) when fired at a sufficiently high temperature. The processes used in the manufacture of ceramics depend on these properties of clays. Different clays are chosen for the different products required, and they are often blended to give the required properties. Before a clay can be used, impurities may have to be removed, for example by froth flotation. Chemical purification is needed for high purity materials such as aluminum oxide and titanium(IV) oxide.

Raw materials used in the manufacture of ceramics are clay, sand and feldspar ...
... with fluxing agents to lower the vitrification temperature ...
... and refractory agents to increase temperature resistance.

To the three main ingredients (clay, sand and feldspar), **fluxing agents** are added to lower the vitrification temperature (glass making temperature). Examples are sodium borate (borax), sodium carbonate, potassium carbonate, calcium fluoride and sodium aluminium fluoride (cryolite). Special **refractory ingredients** are added to increase

Ceramics are made by grinding the components to fine powders, mixing, shaping, and heating the mixture to the firing temperature.

temperature resistance; examples are aluminium oxide, calcium carbonate, calcium oxide, titanium(IV) oxide, aluminium silicate and silicon carbide (carborundum).

All ceramics are made by grinding the components to fine powders, mixing, shaping and heating the mixture to the firing temperature. This may be 700 °C for some overglazes or as high as 2000 °C for many vitreous materials. Firing takes place under high pressure in an electric kiln. The reactions which take place are:

- dehydration (loss of water of crystallisation) at 150–650 °C
- decomposition of e.g. $CaCO_3$ at 600–900 °C
- oxidation of e.g. organic matter at 350–900 °C
- silicate formation at 900 °C and higher.

Many reactions take place. The ceramic product includes aluminium silicate and silica. On heating, ceramics undergo vitrification . . .

. . . giving them a translucent appearance.

A main constituent of clay is an aluminium silicate, kaolinite, $Al_2O_3 \cdot 2SiO_2 \cdot 2H_2O$. When it is fired, it is converted into another aluminium silicate, mullite, $3Al_2O_3 \cdot 2SiO_2$, and crystalline silica, cristoballite, SiO_2. Fluxes lower the temperature of formation of mullite. The composition of ceramics is actually more complex than this: calcium, magnesium and alkali metals may be present. The metal ions and the fluxing agents become part of the vitreous phase of the ceramic. All ceramics undergo vitrification on heating. The extent of vitrification depends on the proportions of refractory oxides and fluxing oxides in the composition and on the temperature and time of heating. The vitreous phase gives some ceramics an attractive appearance, e.g. the translucency of chinaware. Even in refractory ceramics, which are to be used in furnaces etc., some vitrification is an asset because it bonds components together, but extensive vitrification destroys refractory properties. All ceramics consist of two parts: a vitreous matrix and crystals, of which mullite and cristoballite are the most important. Ceramics can be listed according to the degree of vitrification [Table 3.2A].

Refractory ceramics are heat-resistant.

Ceramics	Proportion of fluxes	Temperature of firing	Vitrification
Whitewares	Varying	Moderate	Little
Heavy clay products	Large	High	Little
Refractories	Small	High	Little
Enamels	Large	Moderate	Complete
Glasses	Moderate	High	Complete

TABLE 3.2A

3.2.1 KILNS

The vitrification of ceramic products is carried out in **kilns**. These may be operated continuously or in batch operations. The continuous tunnel kilns have the advantages that the labour costs are lower, fuel efficiency is greater, the processing time cycle is shorter and operating control is better. Gas, coal and oil are the most economic fuels and are the most commonly used, although electricity is used in some cases.

3.3 PROCESSING

Often the processing of a ceramic entails compressing and fusing particles of a powder, e.g. aluminium oxide.

An example of processing a ceramic is the manufacture of a spark plug from aluminium oxide. Aluminium oxide powder of less than 5 μm particle size is placed in a mould and compressed. A small amount of binder is added to maintain strength and shape. The plug is placed in a furnace and heated to about 1800 °C. This causes the object to sinter; that is, the individual particles fuse together. Some shrinkage results.

Injection moulding is used for some objects. Ceramic powder is mixed with a polymer, and shaped as shown in Figure 4.15A.

Injection moulding and slip casting may be used.

Slip casting is another method. Ceramic powder is mixed with a fluid to form a slurry and poured into a porous mould of e.g. plaster of Paris. Slip casting is the method used to make pottery, e.g. plates and saucers.

CHECKPOINT 3.3

1. What is a ceramic? Give three examples of (i) traditional ceramics and (ii) modern engineering ceramics.

2. (*a*) List the mechanical properties of ceramics.

(*b*) Give examples of applications in which ceramics are (i) strong, (ii) weak.

(*c*) What can be done to make a ceramic less brittle?

3. Materials widely used in the manufacture of ceramics are clay, sand and feldspar. List four elements likely to be present in (*a*) a clay, (*b*) a feldspar.

4. What are the reasons for adding (*a*) a fluxing agent, (*b*) a refractory agent in the manufacture of a ceramic? Give one example of each.

5. Name the two phases which make up a ceramic.

3.4 USES OF CERAMICS

3.4.1 WHITEWARE

Whiteware is made by heating selected clays in a kiln with fluxes and then glazing.

Whiteware is a term for ceramic products which are usually white and of fine texture. Selected grades of clay are heated in a kiln with varying amounts of fluxes to a moderately high temperature (1200–1500 °C). The varying amounts of fluxes lead to varying degrees of vitrification in whitewares, from earthenware to vitrified china. Whiteware is often glazed; this is particularly important for tableware. A glaze is a thin layer of glass melted on to the surface. Whiteware includes earthenware, chinaware, porcelain, sanitary ware, stoneware and floor tiles.

3.4.2 STRUCTURAL CLAY PRODUCTS

Structural clay products, e.g. building bricks, pipes and tiles, are made from common clays.

Structural clay products are low-cost but very durable products. They include building brick, face brick, terracotta, sewer pipe and drain tile. They are manufactured from the cheapest of common clays and may or may not be glazed. The clays generally contain sufficient impurities to act as fluxes. A 'salt glaze' may be supplied by throwing salt into the kiln fire: the salt vaporises and coats the ceramic.

3.4.3 REFRACTORY CERAMICS

Refractory ceramics are used in furnaces.

The function of **refractory ceramics** is to withstand the thermal, chemical and physical effects encountered in furnaces. They can be used at much higher temperatures than metals [see § 3.1]. Fluxes are needed to bind together the particles of the refractory ceramics, but these are kept to a minimum to reduce vitrification. Fire clay bricks are the most widely used of all ceramics. Some have a very high silicon(IV) oxide content, and others have a very high aluminium oxide content.

3.4.4   SPECIALISED CERAMIC PRODUCTS

New uses are being found for ceramics all the time. They are used in spacecraft [see § 3.1 Box], in medicine [see Box below], in engineering, e.g. in cutting tools, engine parts and turbine blades, and in electronic components [see below]. Many ceramic composites find important uses [see § 5.1].

Cermets are composite materials made by combining metals and ceramics [see § 5.1]. They combine the tensile strength of metals and the thermal insulation properties of ceramics. Such materials can be used at high temperatures; they are employed for aerospace hardware, e.g. heat shields and rocket nozzles.

New uses for ceramics are found in …
… spacecraft, medicine, engineering …
… replacement of damaged bones.
Cermets are composites of ceramics and metals.

Ceramic implants

In elderly people joints may become worn and painful, and the replacement of these joints by synthetic materials is a common surgical operation. Traditionally, metals and polymers have been used for implants, the prosthetic devices used to replace hip joints, knees and other defective bones. In recent years, dozens of ceramics have been investigated as alternative implant materials. Their resistance to wear and erosion is superior to that of other materials. This means that there is less risk of a prosthesis wearing out and the patient needing a second operation. Surgeons are attracted by the high specific modulus of ceramics, which means that a lighter device can be used, thus increasing the patient's mobility. Another advantage is that ceramics can be made porous, thus allowing regenerating bone to grow into and bond with the ceramic implant. Aluminium oxide and silicon nitride and a silicon oxide based complex called 'bioglass' are in use as ceramic implants [§§ 6.8.2, 6.84]

3.4.5   MAGNETIC PROPERTIES

Some ceramics have magnetic properties.

Ferromagnetic ceramics have important applications. Barium titanate, $BaTiO_3$, is a common example. Ferromagnetic ceramics are used in television sets, computers, magnetic switches, transformers, recorders and memory devices.

3.4.6   ELECTRONIC CERAMICS

Ceramics are used as electronic components …
… from insulators to conductors.

Electronic components make up over 65% of the ceramics market. Ceramics offer a wide range of properties from insulators to conductors. Some ceramics are used as insulators, e.g. in power transmission lines, and others are used as conductors in electronic chip packages and in capacitor materials. Ceramics can be made semiconducting and even superconducting by suitable alloying.

3.4.7   SUPERCONDUCTORS

Some ceramics are superconductors.

Superconductors are substances that conduct electricity with zero resistance [see § 1.13.4]. Hopes of discovering a superconductor which will function at room temperature are centred on ceramic conductors.

3.4.8 VITREOUS ENAMEL

Vitreous enamel or porcelain is applied to metals. It has the virtues of beauty and durability.

Vitreous enamel or **porcelain** is a ceramic mixture containing a high proportion of fluxes. The mixture is applied cold and fused to a metal at a moderate temperature. Complete vitrification takes place. Since ancient times, enamel has been applied to copper, silver and gold to make objects which are valued for their beauty. Now enamel has acquired commercial value because it gives products which are durable, easy to clean and resistant to corrosion. Vitreous enamel is used for plumbing fixtures, kitchen equipment and industrial equipment. Glass-enamelled steel is used in chemical plants.

3.4.9 ENGINEERING COMPONENTS

Ceramic coatings are applied to engine components in vehicles . . .
. . . to improve resistance to wear and tear . . .
. . . and enable engines to run at higher temperatures.

A ceramic coating on cast iron improves resistance to wear and tear and heat and therefore extends the life of engine parts. Ceramic bearings can operate at high speeds without lubricants. Research is going on into the use of ceramics to allow diesel engines and jet engines to run at higher temperatures. The higher the temperature at which an engine operates, the more efficiently it runs and the less fuel it consumes.

Gas turbine blades made of ceramics, e.g. silicon nitride, can run at higher temperatures than alloys.

Machine tools made of ceramics can rotate twice as fast as metal tools without deforming or wearing out. *Sialon* – a ceramic made of silicon, aluminium, oxygen and nitrogen – is as hard as diamond, as strong as steel and as low in density as aluminium.

CHECKPOINT 3.4

1. What is the difference between the ceramics chosen for use in (*a*) cups and saucers, (*b*) building bricks and (*c*) furnace linings?

2. What properties of ceramics make them important in spacecraft?

3. What types of functions are filled by ceramics in electrical circuits?

4. Why are ceramics being used for the replacement of damaged bones?

5. Give three examples of the choice of a ceramic over a metal and explain why a ceramic is better for the purpose.

3.5 STRUCTURE

The bonding in ceramics is ionic or covalent or intermediate between the two.
Ceramics have a crystalline structure . . .
. . . with directed bonds making the structure rigid . . .
. . . so that ceramics are hard, tough, of high melting temperature and of low ductility.

Ceramics consist of a regular arrangement of atoms. The repeating structural units are atoms or ions or covalently bonded molecules arranged in a regular three-dimensional structure. The bonding may be covalent, e.g. diamond and silicon(IV) oxide, ionic, e.g. magnesium oxide, or intermediate between pure ionic and pure covalent, e.g. aluminium oxide. Ceramics include glasses, which have a solid structure in which regular crystallinity has been disordered.

The bonds between atoms in a ceramic are more rigid than in metals so that plastic deformation takes place with difficulty. Ceramics are therefore usually rather brittle. Metals and ceramics represent two extremes of crystalline behaviour, from metals with good ductility and high toughness to ceramics with poor ductility and high toughness. Most ceramics are much harder than metals and have higher melting temperatures. This is due to the powerful directional bonding in these materials.

3.5.1   SILICA

Silicon(IV) oxide, silica, SiO_2, is the basis of a large variety of ceramics. A silicon atom can form covalent bonds with four oxygen atoms to form the unit SiO_4^{4-} with a tetrahedral distribution of bonds [see Figure 3.5A]. The SiO_4^{4-} anion occurs in minerals such as zircon, $ZrSiO_4$, and topaz, $Al_2SiO_4F_2(OH)_2$.

FIGURE 3.5A   The SiO_4^{4-} Tetrahedral Ion ($\bullet$ = silicon, $\circ$ = oxygen)

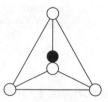

Each of the four oxygen atoms in SiO_4^{4-} has an unshared electron which it can use to combine with other SiO_4^{4-} units. Anions such as $Si_2O_7^{6-}$, $Si_3O_9^{6-}$ and $Si_6O_{18}^{12-}$ occur in many minerals [see Figure 3.5B].

FIGURE 3.5B
The Anion $Si_3O_9^{6-}$

The SiO_4^{4-} anion has a tetrahedral distribution of bonds. It is the basis of a number of minerals. Other silicate ions are $Si_2O_7^{6-}$, $Si_3O_9^{6-}$ and $Si_6O_{18}^{12-}$

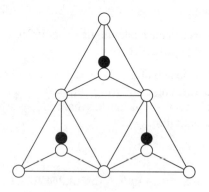

3.5.2   CHAIN STRUCTURES

Silicates contain long chains of linked tetrahedra, $(SiO_3)_n^{2n-}$ Metal cations are associated with these anions to balance the charges.

In addition to silicates which contain discrete anions, there are silicates made of long chains of linked tetrahedra. Some minerals have single silicate strands of formula $(SiO_3)_n^{2n-}$, which are bonded to metal ions which balance the negative charges [see Figure 3.5C]. Asbestos has the double-stranded structure shown in Figure 3.5D. The double strands are bonded to other double strands by ionic bonds to the Na^+, Fe^{2+} and Fe^{3+} ions packed round them. The double strands can be separated from other double strands with less force than is required to break covalent bonds in the chain, and asbestos has therefore a fibrous texture.

FIGURE 3.5C   Chains of Silicate Tetrahedra

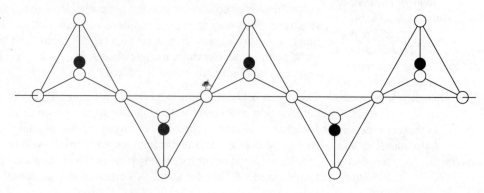

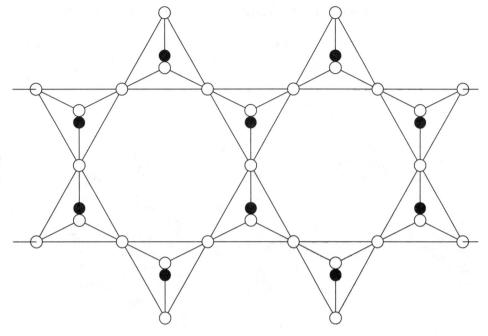

*Double-stranded
chains are found in some
minerals, e.g. asbestos.*

Aluminium atoms replace silicon atoms in many silicate minerals. For every
aluminium atom that replaces a silicon atom, an ion, e.g. Na^+ is needed to balance the
charge.

3.5.3 SHEET STRUCTURES

*Sheets of combined
$SiO_4{}^{4-}$ units are held
together . . .
. . . by van der Waals forces
in e.g. talc . . .
. . . and by ionic bonds to
metal cations in e.g. mica.
Clay minerals are silicates
with a sheet structure, in
which some silicon is
replaced by aluminium.*

Silicate tetrahedra combine to form silicates with sheet structures (similar to
Figure 3.5E). Talc has this structure, with the silicate sheets held together by van der
Waals forces. This accounts for the soft texture of talc, which is at the bottom of the
Mohs hardness scale [see § 1.7]. Mica has a similar structure, but the silicate sheets are
bonded through the formation of ionic bonds to layers of metal cations between the
sheets [Figure 3.5D]. This structure gives mica its flaky character. Clay minerals also
are silicates with the sheet structure shown in Figure 3.5E, but in clays one quarter of
the silicon atoms have been replaced by aluminium atoms. Replacing silicon, with
oxidation state +4, by aluminium, with oxidation state +3, means that additional
positive charges, e.g. Na^+ ions, are needed to balance the charges on the oxygen
atoms. Layers of cations, Na^+ and Al^{3+}, between layers hold the layers together by
ionic bonding. These layer structures have large inner surfaces which can absorb large
amounts of water.

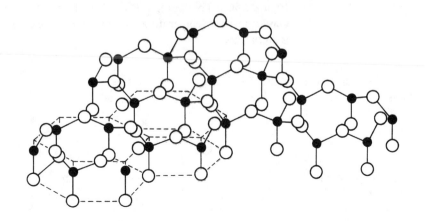

3.5.4 THREE-DIMENSIONAL STRUCTURES

Quartz and other crystalline forms of silicon(IV) oxide have a three-dimensional network of SiO_4 units, in which all four oxygen atoms of the silicate tetrahedra are shared with other Si atoms [see Figure 3.5F]. With each silicon atom bonded to four oxygen atoms and each oxygen atom bonded to two silicon atoms, the formula is $(SiO_2)_n$.

FIGURE 3.5F
The Three-Dimensional Network of Silicate Tetrahedra in Quartz, SiO_2

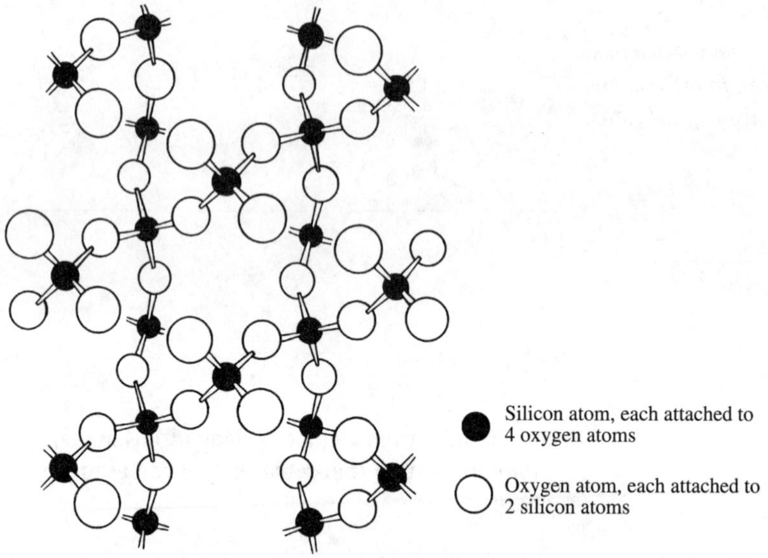

Quartz has a three-dimensional network of SiO_4 units combined to give the formula SiO_2.

● Silicon atom, each attached to 4 oxygen atoms

○ Oxygen atom, each attached to 2 silicon atoms

In quartz, all the tetrahedral structures have silicon atoms, but in other minerals up to half the silicon atoms can be replaced by aluminium atoms. An example is feldspar, $KAlSi_3O_8$ [see Mohs hardness scale, § 1.7]. The chief component of the Earth's crust is granite, which is a mixture of mica, feldspar and quartz.

3.5.5 GLASSES

When molten silica is cooled very slowly, it crystallises at the freezing temperature. If molten silica is cooled more rapidly, it is unable to organise all the atoms into the ordered arrangement required for a crystal and it solidifies as a disordered arrangement which is called a **glass** [see Figure 3.5G]. Glasses are amorphous, disordered non-crystalline substances composed of linked silicate chains [see Figure 3.5G]. The temperature at which molten silica turns into a glass is called the **glass transition temperature**. It depends on the rate of cooling of the molten material.

FIGURE 3.5G
A Glass – a disordered chain of silicate tetrahedra

When molten silica is cooled very slowly, it crystallises ...

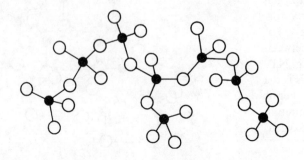

A glass can be considered to be a supercooled liquid. When a cooled liquid solidifies, there is a sudden change in density and a release of the enthalpy of freezing when the liquid changes state. These sudden changes do not happen with a glass. A glass contracts when its temperature falls as the atoms rearrange and pack more closely together [see Figure 3.5H]. There is no abrupt change as there is when a liquid solidifies: a glass just goes on contracting as if it were a liquid. Eventually it reaches a temperature when the closer packing of the atoms has to cease. The change in volume is smaller than that which occurs when a liquid crystallises. The temperature at which the contraction occurs is the **glass transition temperature**.

but when it is cooled more rapidly to the glass transition temperature it solidifies as a glass. The glass transition temperature depends on the rate of cooling.

A glass will sometimes crystallise if given sufficient time. The process is called **devitrification** and may take several hundred years. It is sometimes found in very old stained glass windows and results in an increase in brittleness.

FIGURE 3.5H
The Glass Transition
Temperature T_g

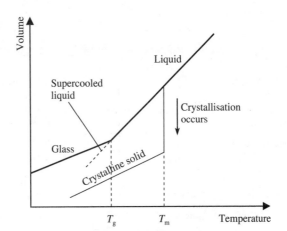

3.5.6 BRITTLENESS

Metals have slip planes [see § 2.7]. In an ionic crystal, there are fewer slip planes than in a metal because slip cannot be allowed to bring cations alongside other cations and anions alongside other anions [see Figure 3.5I].

FIGURE 3.5I
Slip in an Ionic Crystal

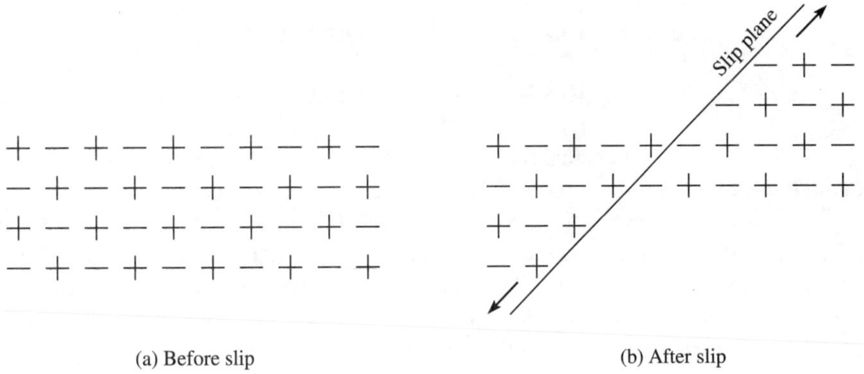

(a) Before slip

(b) After slip

In crystalline ceramics, slip cannot occur, and these materials are brittle. In ceramics with a layer structure, layers can slide over one another ... ... and the materials are soft and flaky.

Since few planes in an ionic crystal are able to slip, crystalline ceramics are brittle materials. The picture is different in ceramics such as talc which have a layered structure with strong covalent bonds between the atoms in a layer and weak van der Waals forces between layers. Structures of this kind allow one plane to slip over another. A similar situation obtains in ceramics such as mica and clays which have a layered structure in which sheets of covalently bonded silicate tetrahedra alternate with layers of cations [Figures 3.5D,E]. Again, one layer can slide over another.

Glasses do not have slip planes ... The disordered arrangement of atoms in glasses means that they do not have slip planes like crystalline materials. The bonding of atoms in glasses is three-dimensional and the movement of an atom or group of atoms through a glass is not possible; this is why glasses are brittle.

... and are brittle.

3.5.7 TRANSPARENCY

Glasses are amorphous and transparent. Crystalline ceramics are opaque. Glass is transparent. The reason is that glass is amorphous: the arrangement of atoms is disordered and provides no reflecting surfaces, such as the grain boundaries which make metals opaque. Crystalline ceramics, on the other hand, have regularly spaced layers of atoms which reflect light, and these ceramics are opaque.

CHECKPOINT 3.5

1. (a) Show the electron structure of SiO_4^{4-} by means of a 'dot and cross' diagram.

(b) Show how this unit can act as the building block of silicate minerals.

2. (a) Why are glasses classified as ceramics?

(b) What is the glass transition temperature?

3. Describe how silicon(IV) oxide can form the basis of a range of ceramics with different properties.

4. (a) How do glasses differ from quartz in properties?

(b) How are these differences related to the difference in structure?

5. What behaviour is observed when a force sufficient to deform the object acts on (a) a metal object, (b) a crystal of a salt, (c) a lump of clay, (d) a block of glass? Explain how the difference in behaviour relates to the difference in structure.

6. Why is glass transparent and china opaque?

7. Explain why ceramics are (a) harder than metals, (b) more brittle than metals, (c) generally poor electrical conductors.

8. (a) What is a 'slip plane'?

(b) Compare metals and ceramics with respect to their ability to slip.

3.6 CEMENT AND CONCRETE

3.6.1 PORTLAND CEMENT

Roads, bridges, dams and tunnels, public buildings and homes, reinforced concrete girders and walls – everywhere we look we witness the importance of cement and concrete in our lifestyle. This importance dates back to ancient times. The Egyptians used a cement in the construction of the Pyramids, and the Greeks and Romans mixed volcanic ash and lime to make cement. In 1823, a Briton, Joseph Aspdin, patented a cement which he had made by strongly heating limestone that contained silica. He called it **Portland cement** because concrete made from it resembled the popular building stone from Portland. The name Portland cement is still given to any material made by strongly heating a mixture of limestone and clay or similar materials. **Concrete** is artificial stone made from a mixture of cement, water and fine and coarse **aggregate** (usually sand and coarse rock).

Cement has been known since ancient times. Portland cement has been a popular building material since 1823. Concrete is made from cement, water and aggregate.

Raw materials used in the manufacture of cement are: a source of calcium, and a source of silica. The manufacture of cement requires two raw materials, one rich in calcium, e.g. limestone, chalk, aragonite, and one rich in silica, e.g. clay, shale, pumice, volcanic material. **Argillaceous limestone**, called **cement rock**, is used as a source of both. Sand, sandstone, quartz, waste bauxite, blast furnace slag and iron ore are sometimes added. Calcium sulphate (4–5%) is added to regulate the setting time of the cement. The raw materials are finely ground and mixed. They are heated in a rotary kiln to form

The raw materials are ground, mixed and heated in a rotary kiln to form cement clinker. It consists of a silicate structure associated with the ions of calcium, aluminium, iron(II) and magnesium.

cement clinker. A number of reactions take place, including the decomposition of calcium carbonate and the reaction between calcium oxide and silica. Towards the end of the process, at 1250 °C, fusion occurs to form a calcium silicate of formula Ca_3SiO_5. This is the component that gives cement most of its strength. Other constituents include other calcium silicates, calcium aluminates and magnesium oxide. Minute quantities of resinous materials that have pockets which can trap air are added to cement. The air pockets help hardened concrete to withstand repeated freezing and thawing without cracking.

Cement clinker is made by both wet and dry processes. In the original wet process, the mixture of ground raw materials is mixed with water to form a slurry, which is fed into the kiln. In the dry process, the raw materials are ground, thoroughly mixed and fed dry into the kiln. The dry process is taking over because of the saving in heat. The kiln is 45 m long in the dry process or 90–180 m long in the wet process. The diameter is 2.5–6 m, and the kiln rotates at 1/2 to 2 revolutions per minute. The kilns are slightly inclined, and materials travel through in 1–3 hours. For the refractory lining of the kiln, high-alumina bricks and high-magnesia bricks are used. The product, clinker, which leaves the kiln is hard grains of 3–20 mm in diameter. The cement industry uses a large amount of energy. Research is being carried out to find ways of making it more energy efficient, for example by finding fluxes which will reduce the temperature needed for the formation of clinker.

FIGURE 3.6A
Flow Diagram for
Cement Manufacture

Cement clinker is made by both wet and dry processes. A long rotating kiln at a slight angle to the horizontal is used. The exhaust gases are subjected to electrostatic precipitation.

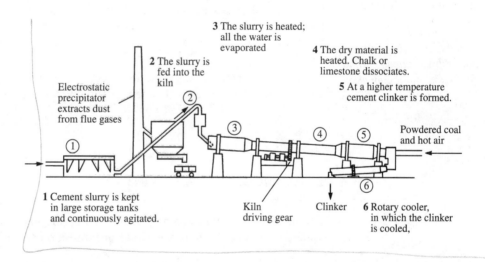

3 The slurry is heated; all the water is evaporated

4 The dry material is heated. Chalk or limestone dissociates.

2 The slurry is fed into the kiln

5 At a higher temperature cement clinker is formed.

Electrostatic precipitator extracts dust from flue gases

Powdered coal and hot air

1 Cement slurry is kept in large storage tanks and continuously agitated.

Kiln driving gear

Clinker

6 Rotary cooler, in which the clinker is cooled,

3.6.2 SPECIAL APPLICATIONS

Concrete based on cement is being used in wider applications each year. For different uses, concretes are available which are low-temperature-hardening, quick-hardening, slow-hardening, low-density, high-density and steel-reinforced.

Reinforced concrete beams have great strength [see § 5.5] and are used in construction.

High-alumina cements are used when resistance to sea water is important.

Silicate cements are used when resistance to acid attack is important.

Sulphur cements are used for joining tiles and cast iron pipes.

Modifications of concrete are available to suit different applications. Composite materials of concrete and polymer give improved characteristics for specific applications.

Polymer concretes are composite materials, consisting of aggregate and resins, e.g. epoxy resin, methyl 2-methylpropenoate and polyester; they do not contain Portland cement. Each different resin gives a certain characteristic to the concrete, e.g. rapid curing, resistance to corrosion and high compressive strength. Polymer concretes are much more expensive than Portland cement concrete.

3.6.3 TOUGHER CONCRETE

The low tensile strength of unreinforced concrete is due to its porous nature. Various methods have been tried to reduce porosity.

Concrete is one of the most widely used structural materials. It is relatively cheap and has high compressive strength. Research and development have aimed to improve the tensile strength and toughness of concrete. Unreinforced concrete is unsuitable for many applications where good tensile strength is required. For such applications, metals are used. Research has shown that the low tensile strength of concrete is due to the presence of microscopic pores. The larger pores make it difficult for particles to pack together, with a resultant loss of strength. If porosity can be eliminated from cement, it gains a tensile strength similar to that of mild steel. Various methods have been tried to reduce porosity. The cement can be subjected to **vibrational compaction** in a mechanical shaker to pack the particles together. Another method is to add pore-filling materials, e.g. sulphur and resins. Water-soluble organic polymers can be added to the cement–water mixture. The polymer–water gel effectively fills any excess space left between the cement particles. The resulting material is processed by one of the methods used to mould plastics, e.g. extrusion or rolling. Once the material is formed into the desired shape, it hardens. As a result of the filler and the type of process used, the final material consists of polymer-bonded close-packed cement grains containing little or no porosity.

A mechanical shaker can be used to pack particles of cement together. The pores may be filled with sulphur and with resins. Water-soluble polymers can be added to the cement–water mixture.

Other filler materials used in polymer concretes are glass fibres, silicon carbide fibres and aluminium oxide particles or fibres.

Other filler materials, such as glass fibres, silicon carbide, alumina particles or fibres have been used in the polymer concretes. Materials with a range of properties can be obtained.

A **superplastic cement is** based on fine silicon nitride and silicon carbide grains in a glue of oxide. This material is more plastic than most metals.

3.6.4 CERAMICS SUMMARISED

Ceramics are crystalline compounds of oxygen, carbon or nitrogen with other elements, e.g. silicon, aluminium and some transition metals.

The bonding in ceramics may be ionic or covalent. The bonds are directed in space, and the structure is therefore more rigid than a metallic structure. This structure makes ceramics:

- crystalline
- hard (able to cut steel and glass)
- of high melting temperature
- poor conductors of heat and electricity because they lack free electrons
- usually opaque because light is reflected by grain boundaries in the crystal structure
- not ductile or malleable or plastic, except at high temperature
- rather brittle because the bonds are rigid, making it difficult to change the shape without breaking

The properties of ceramics are summarised.

- resistant to corrosion.

1. Review the reasons why concrete is an important building material. Discuss the availability and cost of the necessary raw materials, the strength of concrete and the variety of uses to which it is put, including special varieties of concrete.

2. What methods can be employed to increase the strength of concrete?

3. Compare concrete and steel as possible building materials for use in an office block (*a*) for structural beams, (*b*) for the floor, (*c*) for the walls.

4. What is the advantage of a ceramic over (i) an alloy and (ii) a hard plastic for use as (*a*) a false tooth, (*b*) a knee implant, (*c*) an artificial middle ear, (*d*) a space shuttle tile?

5. The original cooking pots were made of ceramics. How did Bronze Age people find that bronze vessels compared with them?

3.7 THE HISTORY OF GLASS

Glass has been made and used since ancient times.

Glass has been known for centuries. The Egyptians made glass imitation jewels in 5000 BC. Window glass was recorded in AD 290, and a method of making it on a large scale was invented in the twelfth century, but it did not come into wider use until the fifteenth century. Venice was the medieval centre of the glass industry, and Britain did not make glass until the sixteenth century. A method of making **plate glass** was invented as long ago as the twelfth century. A rod of glass was softened by heat, and a 'gob' of glass was cut off and blown into a cylinder. Then the ends of the cylinder were cut off, the cylinder was cut open, then softened and rolled into the form of a plate of glass. This laborious method was used for centuries. The process was completely manual and rule-of-thumb. The only chemical improvements made in three centuries were purification of the batch materials. In 1914 the Fourcault process was invented in Belgium. The process carried out these operations continuously and produced a plate glass with fewer imperfections and distortions. The quality of window glass suddenly improved. Scientists and engineers worked on the new process to reduce the cost of all the grinding and polishing that were necessary to obtain flat surfaces on the glass.

A major advance came in the 1960s when a British firm, Pilkington, perfected the float glass process. The raw materials are fed into a furnace. Molten glass is channelled out of the furnace onto a bath of molten tin. The surface of the liquid tin is perfectly smooth. The molten glass is conducted along the surface of the tin in a non-oxidising atmosphere at a closely controlled temperature. All irregularities in the glass

FIGURE 3.7A
Flowchart of the Float
Glass Process

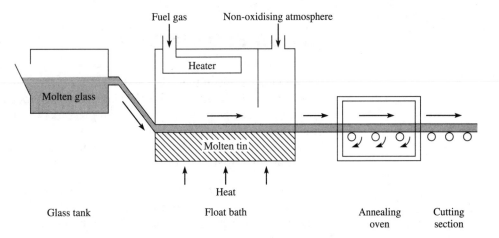

A major development in recent times is the invention of the float glass process for making plate glass.

are melted out, and a plate of glass with both surfaces flat and parallel is produced. A 2.5 m wide glass ribbon comes off the production line and is cut to the sizes and shapes required. Why was tin the chosen material for the bath? Tin has a density greater than that of glass. It is molten at under 600 °C but has a high boiling temperature, and it does not react chemically with glass.

FIGURE 3.7B
The Float Glass Process

Glasses are ceramics ...
... they are supercooled liquids ...
... with a random structure ...
... made from sand, sodium carbonate and lime (calcium oxide).

Glasses are ceramics [see the definition in § 3.1]. A glass is a supercooled liquid with no definite melting temperature, a random structure [see § 3.5] and a viscosity which is high enough to prevent crystallisation. Glass is made by melting together sand, sodium carbonate, lime and other compounds. Glass has many uses because of its transparency, its high resistance to chemical attack, its electrical resistivity and its ability to contain gas at a reduced pressure without imploding. Glass is a brittle material, with compressive strength much greater than its tensile strength. Research workers have found methods of strengthening glass to adapt it to new uses. About 800 different types of glass are made.

Glass production is divided into:

Glass is transparent, resists chemical attack, has high electrical resistivity, can withstand compression, but is brittle.

- flat glass
- pressed and blown glass
- glass containers.

3.8 GLASS BLOWING

Glass blowing is one of the oldest crafts. Until a century ago, it depended on human lungs and human hands to shape the glass. Now the production of bottles is a mechanical casting operation that uses air pressure [see Figure 3.8A]. A gob of molten glass is delivered into a heated mould, and air is injected to form a bubble of glass and force it into the shape of the mould. Different machines make light bulbs and television tubes by similar techniques.

All glass objects must be **annealed** to reduce strain, whether they are hand-made or machine-made. Two operations are involved. First the object is held above a critical temperature long enough for plastic flow to occur and reduce internal strain. Next the

FIGURE 3.8A
Making a Glass Bottle

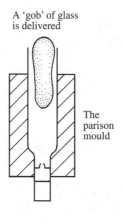

A 'gob' of glass is delivered

The parison mould

1

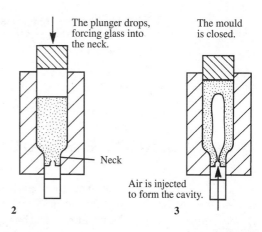

The plunger drops, forcing glass into the neck.

The mould is closed.

Neck

Air is injected to form the cavity.

2   **3**

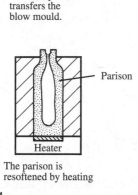

The parison mould inverts, opens, and transfers the blow mould.

Parison

Heater

The parison is resoftened by heating

4

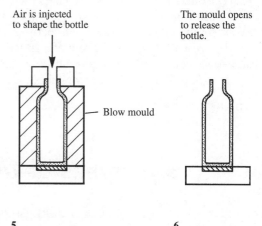

Air is injected to shape the bottle

Blow mould

The mould opens to release the bottle.

5   **6**

Glass objects are made by glass blowing . . .
. . . either by hand or by means of a machine which uses air pressure . . .
. . . and then annealed to reduce strain.

object is cooled to room temperature slowly enough to hold the strain at this reduced level. This is done in an annealing oven, a heated chamber in which the rate of cooling can be carefully controlled. After annealing, the glass object goes through one or more finishing operations, such as cleaning, grinding, polishing, cutting and sandblasting.

3.9 RAW MATERIALS

The raw materials needed for glass manufacture are sand, lime and sodium carbonate, just as they were two thousand years ago. Other raw materials, which are included in small amounts, may have major effects on the nature of the glass. The important factor in making glass is the viscosity of the molten mixture, and this is related to the composition. The raw materials are:

The raw materials used in glass manufacture are calcium oxide (from lime kilns), silicon(IV) oxide (as sand) and sodium carbonate (from the ammonia–soda process) . . .
. . . and other materials . . .
. . . and cullet (crushed glass).

- sand, which should be almost pure quartz, SiO_2, with a very low iron content
- sodium carbonate, a product of the ICI ammonia–soda process [see *ALC*, § 18.8.1]
- calcium oxide, made by heating limestone in lime kilns [see *ALC*, §§ 18.5.5, 18.5.6]
- magnesium oxide and calcium oxide, made by heating dolomite, $CaCO_3 \cdot MgCO_3$
- feldspar, $M_2O \cdot Al_2O_3 \cdot 6SiO_2$ (where M = Na or K or both) is mined
- sodium borate (borax) $Na_4B_2O_7$
- sodium sulphate
- cullet, crushed glass from imperfect articles and recycled glass

The chemical reactions which take place result in the formation of sodium silicates, of formulae $Na_2O \cdot aSiO_2$, and calcium silicates of formulae $CaO \cdot bSiO_2$ (where a and b are not integers).

3.10 THE STRUCTURE OF GLASS

Glass is amorphous ...
... and transparent ...
... brittle – largely because
of surface defects and
scratches – ...
... except for glass fibres
which are not brittle
because of the absence of
surface defects.

The structure of glass was discussed in § 3.5. In the formation of plate glass, the transition from liquid to amorphous solid is achieved by moderate rates of cooling. Non-uniform cooling can result in stress in the glass which results in fracture if the glass is loaded. The **brittleness** of glass comes mainly from surface defects and scratches. Glass breaks easily if a **scratch** is first made on the surface [see § 1.10.1]. When glass is loaded, the stress at the root of the scratch is greater than the strength of the bonds between atoms, and the glass breaks along the scratch. In metals, stress at the root of a surface crack can be countered by dislocation movements. The reason why **glass fibres** can be bent without breaking is that they are almost entirely free of surface scratches.

3.11 TYPES OF GLASS

SILICA GLASS, OR FUSED SILICA, OR VITREOUS SILICA

This is made by melting quartz or pure sand or by the pyrolysis of silicon tetrachloride. Silica glass finds applications, e.g. in laboratory apparatus, because of its high softening temperature and because of its transparency to ultraviolet light.

Types of glass include:
silica glass – used in
laboratory apparatus ...
alkali silicates,
e.g. water glass ...

ALKALI SILICATES

These are made by melting together sand and sodium carbonate to form sodium silicates. A solution of sodium silicate called **water glass** is used as an adhesive.

SODA-LIME GLASS

This makes up 95% of all glass manufactured. It is used for containers of all kinds, as float glass, vehicle windows and tableware. Float glass has the composition 70–74% SiO_2, 8–13% CaO, 13–18% Na_2O. The melting point is relatively low. The glass is sufficiently viscous that it does not devitrify but not too viscous to be workable at a convenient temperature. The composition of the glass has not changed for centuries, but methods of manufacture have changed by the substitution of instrument-controlled machines for human operators. Similarly with glass containers, developments have been along the lines of instrumentation.

... soda-lime glass – 95%
of all glass manufactured,
lead glass – cut glass, light
bulbs, neon signs ...

LEAD GLASS

When lead oxide is substituted for calcium oxide in the melt, lead glass is formed. Lead glasses have high values of refractive index. They are used for making cut glass objects, for electric light bulbs and for neon signs.

BOROSILICATE GLASS

A high proportion of boron oxide gives a borosilicate glass. These have a low coefficient of expansion, high resistance to shock, high chemical stability and high electric resistance. Laboratory glassware made from this glass is sold under the trade name **Pyrex**. Borosilicate glass is also used for electric insulators, pipelines and telescope lenses.

COLOURED GLASS AND COATED GLASS

Coloured glasses have been used for centuries for decoration. Today there are scientific and technical uses for coloured glasses. Oxides of transition metals, especially titanium, vanadium, chromium, manganese, iron, cobalt, nickel and copper are included in the melt to colour glass. Coated glasses are made by depositing transparent metallic films on the surface of clear or coloured glass. Buildings can be made with walls of glass that transmit light into the building and also reflect light so that people outside cannot see through. This is done by using films with suitable transmission and reflection characteristics.

SAFETY GLASS

There are two types of safety glass: laminated safety glass and tempered glass. **Laminated glass** consists of two sheets of thin plate glass with a sheet of non-brittle plastic material in between. An adhesive is applied to the glass, and the glass and plastic sheets are pressed together at a moderate temperature. The safety feature is the ability of the plastic to hold fragments of glass if there is a breakage. **Tempered glass** is very strong and tough. It is used for doors and vehicle windows. It is made by heat annealing, which replaces random stresses in glass by a uniform, low-level stress. This glass is very strong in compression and very weak in tension. Annealing is done by heating to a moderate temperature, e.g. 400 °C, just below the softening point of the glass, then quenching in air or oil.

PHOTOCHROMIC SILICATE GLASSES

The property of photochromic glasses is that they darken in light from the visible and ultraviolet spectrum and fade in the dark. The photochromic property is reversible: it can follow thousands of cycles without any deterioration in performance. The glass contains sub-microscopic particles of silver halide, about 5 nm in diameter. These 'colour centres' are embedded in chemically inert glass so that they can neither diffuse away and join to form stable larger silver particles nor undergo the irreversible decomposition of silver halide that takes place in the photographic process. The reversible photochromic process is:

$$AgCl \xrightleftharpoons[\text{light outdoors, night-time}]{\text{light outdoors, daytime}} Ag + Cl$$

... borosilicate glass, e.g. Pyrex®, coloured glasses – which contain transition metal compounds, coated glass – which reflects light ...

... safety glass – laminated and tempered glass, photochromic glass – darkens in sunlight.

3.11.1 GLASS CERAMICS

Glass ceramics are made in a glassy state, then converted largely into a crystalline ceramic by controlled devitrification. The glass is encouraged to crystallise by being held at a temperature in the devitrification range after the addition of some small crystals. The crystals in a glass ceramic are much smaller and more uniform than

Glass ceramics ... are more rigid than glasses ...

those in conventional ceramics. The properties of ceramics produced from glass are closer to those of ceramics than to those of glasses. Glass ceramics are more rigid than glasses and have better thermal and mechanical properties. They may be white or coloured and are usually opaque, glossy, non-porous, more refractory than glasses, less refractory than ceramics. They can be worked by the methods used to shape glass. They are used in guided missiles, electronic devices, cook–serve–freeze dishes and for other uses.

... with better thermal and mechanical properties ... ... and are less refractory than ceramics. ...

3.11.2   GLASS FIBRES

Fibres of glass of 5–10 μm diameter can be drawn into threads or blown into mats to be used as insulation, air filters, etc. The drawn fibres have high tensile strength because of the absence of cracks in the surface [see § 1.10]. They are used to reinforce plastics, e.g. epoxies and polyesters. The composite material [see § 5.3] is made into pipes, tanks, sporting goods, e.g. fishing rods and skis. The glass used for the production of fibres is low in silicon(IV) oxide and high in aluminium oxide and boron oxide. The glass fibres which are used for the reinforcement of concrete should not be attacked and weakened by alkali. The most successful contain 17% of zirconium oxide, which makes these fibres costly.

Glass fibres are used to reinforce plastics ... ... in e.g. fishing rods and skis

FIBRE-OPTICS

Light can flow along glass fibres. Light waves bounce along a glass fibre enclosed in a sleeve of high refractive index. Advanced technology makes it possible to transmit light through fibres over 200 km [see Figure 3.11A].

FIGURE 3.11A
Light Waves Travelling along a Glass Fibre

... and used in fibre optics. Light can flow along glass fibres up to a distance of 200 km.

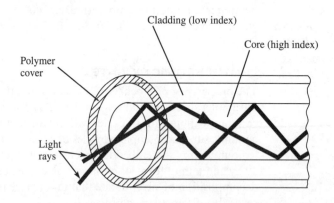

Light can travel along the contours of the glass fibre as it winds its way from one place to another. The fibres are made in such a way that the composition of the glass, and therefore its refractive index, varies from less pure glass at the outside to a very pure fine fibre in the centre [Figure 3.11A]. Light rays are confined to the central micrometre of the fibre.

An advantage of glass fibres over copper conductors in telecommunications is the extremely small diameter of glass fibres compared with copper wires and their greater load-carrying capacity. Since a glass fibre has a diameter of less than one-tenth of a millimetre including cladding, many more glass fibres can be fitted into a cable of a certain size than can copper wires. Also fibre optical carriers are not affected by electromagnetic disturbances. Networks of fibre-optical cables could link individual homes to TV networks. Videophones could become a reality with fibre optics. Transoceanic fibre-optical communications are planned.

Telecommunications systems use glass fibres. A cable of a certain size can carry many more glass fibres than copper wires.

1. Compare the annealing of glass with the annealing of metal [§ 2.10.4].

2. What technique is employed to ensure that plate glass has completely flat surfaces?

3. What changes in the properties of glass are achieved by adding (*a*) lead oxide, (*b*) boron oxide, (*c*) a transition metal oxide, (*d*) microscopic particles of silver halide?

4. Two types of safety glass are laminated glass and tempered glass. Describe the difference between them.

5. What is a glass ceramic? How does it compare with (i) a glass and (ii) a ceramic?

6. What use is made of glass fibres in composite materials?

7. Fibre optics involves glass fibres. (*a*) Why are these fibres so efficient at transmitting light? (*b*) Explain why they are used in telecommunications.

3.12 RECYCLING GLASS

FIGURE 3.12A
Chemicals Used in Glass
Making

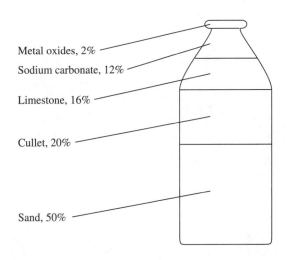

Metal oxides, 2%
Sodium carbonate, 12%
Limestone, 16%
Cullet, 20%
Sand, 50%

Waste glass can be recycled successfully. Bottle Banks collect used glass . . .
. . . saving natural resources and energy . . .
. . . and reducing litter.

Waste glass is known as cullet. It is used in making new glass [see Figure 3.12A]. This saves raw materials and also saves energy because cullet makes the mixture used to make glass melt at a lower temperature. Only 20% of cullet is used in the mixture, but it is possible to use larger amounts if the cullet is available. Bottle banks collect used bottles and jars which are then sold to glass manufacturers. The manufacturers guarantee to the local authority that operates the bottle bank that they will buy the glass at a certain price. Benefits of the scheme are:

- saving in energy used in glass manufacture

- saving in natural resources

- saving in refuse disposal costs

- less glass litter and broken glass in the environment

- a source of income for local authorities.

QUESTIONS ON CHAPTER 3

1. Choose three examples of manufactured ceramic articles.

(*a*) Explain what properties lead to the choice of ceramics for these articles.

(*b*) Explain how these properties arise form the bonding and structures of the ceramics.

2. (*a*) Explain how a solid glass differs structurally from a solid electrolyte.

(*b*) What is meant by 'glass transition temperature'?

(*c*) Explain why the glass transition temperature is lower than the melting temperature of a compound.

3. Choose three examples of manufactured glass articles. Explain what properties of glass make it the material chosen for these articles.

4. (*a*) Sketch the structure of a silicate glass.

(*b*) Give the coordination number of silicon in a silicate glass.

(*c*) Say how the structure of crystalline silicon(IV) oxide differs from that of silicate glass.

(*d*) Describe what happens to the structure of glasses as they cool from the molten state.

(*e*) Explain what is meant by devitrification.

5. (*a*) How would the strength of a block of glass be affected by using hydrofluoric acid to remove surface flaws?

(*b*) A scratch is made in the middle of a glass rod parallel to the ends. The rod can now be broken easily by applying small forces to the ends. Explain why. (See Chapter 1 if necessary.)

6. Review the benefits of recycling glass.

4

POLYMERS

4.1 PLASTICS IN USE

Plastics are all around us. In the sitting room of fifty years ago, the settee and chairs would have been made of natural materials, such as wool, cotton and leather. Today the probability is that they are made of synthetic fibres such as nylons and polyesters. The carpet probably contains a high percentage of nylon or poly(propene), and the curtains contain polyester. The television set, the hi-fi, the video-recorder all have parts made of tough plastics.

In the kitchen, the worktop and table top may be made of thermosetting plastics which are heat-resistant as well as being easy to clean and hygienic. The insides of the refrigerator and freezer are made of plastics. The cupboards contain plastic bowls and buckets and possibly plastic microwave dishes and plates.

FIGURE 4.1A
Plastics in the Kitchen

Our homes contain many plastic materials in a variety of forms.

The bathroom is likely to contain an acrylic plastic bath with plumbing made of poly(propene) pipes.

On the outside of the house, the window-frames, the gutters, the drain-pipes and sewage pipes may well be of PVC. The paint and varnish on the house contain plastics. The car which stands outside the door contains about 100 kg of plastics.

A novel application

As well as replacing traditional materials, plastics find some completely new applications. The plastic poly(propenol) can absorb up to 600 times its own volume of water. An application that makes use of this remarkable property is to promote the growth of seedlings in desert conditions. A handful of plastic granules is placed at the bottom of a hole, and allowed to soak up water. A seedling is planted, the hole is filled, and the plastic releases water gradually at the demand of the seedling. The technique has been tried out successfully on acacia trees in the Sudan and eucalyptus trees in India.

Some plastics have unusual applications.

Plastics are **polymers**, and we shall now look into their chemical composition.

4.2 LONG MOLECULES

A polymer molecule is composed of a large number of repeating units. A polymer is formed by the polymerisation of a monomer.

Some of the plastics mentioned in §4.1 are **biopolymers** and others are **synthetic polymers**. Many of the early synthetic polymers were based on naturally occurring polymers, e.g. celluloid, rayon and vulcanised rubber.

A **polymer** is a substance with molecules built up from many smaller repeating units, which are connected by covalent bonds: —P—P—P—P—P— or —$(P)_n$—. One molecule of polymer is formed by the combination of many molecules of **monomer**, for example, many molecules of the monomer ethene $CH_2{=}CH_2$ polymerise to form one molecule of the polymer poly(ethene), $(CH_2)_n$. The repeat unit, P, cannot exist on its own. P has a reactive group at each end; it is a **bifunctional entity**. The group may be a —CH_2— group as in poly(ethene). There may be an acidic group at one end and a basic group at the other end as in a polypeptide, $H_2N(CHRCONH)_nCHRCO_2H$. Linear polymers are not the only synthetic polymers. The monomers may be **trifunctional** so that polymer chains may become branched and interconnected to form three-dimensional networks. An example is phenol–methanal resin [see §4.12.2]. In a **homopolymer**, all the repeating units are the same; in a **heteropolymer**, they are of two or more kinds.

The repeating unit of a polymer is bifunctional – has two reacting groups – which enable it to form long chains …

… or trifunctional – with three reactive groups – which enable it to form a three-dimensional network of cross-linked chains.

Sulphur, starch, cellulose and proteins are polymers; see *ALC*, §§21.4.2, 31.8.2, 33.17.2.

The degree of polymerisation in the polymer $(P)_n$ is n. The value of n may be small, e.g. 2–20, in which case the polymers are referred to as **oligomers**, and can often polymerise further. In many polymers, n is large and the relative molecular mass may be of the order of 1×10^6. A molecule of a polymer of relative molecular mass 1×10^7 would, if fully extended, have a length of about 1 mm and a diameter of about 0.5 nm. These dimensions would make it the same shape as a piece of spaghetti of length 2 km. In reality, it is shaped more like a miniature ball of cooked spaghetti. A polymer chain is never fully extended; it coils in a random manner to occupy a sphere of about 200 nm in diameter. The configuration of the chain depends on:

Polymer molecules contain a very large number of repeating units … have a high relative molecular mass… can be as long as 1 mm and may be coiled into a ball of 200 nm diameter.
A number of factors influence the configuration of the chain.

● the structure of the chain – whether it consists of rigid aromatic rings or flexible aliphatic carbon chains and the presence or absence of side-chains

● the type of interaction between chains, whether weak dipole–dipole attractions, hydrogen bonds or strong covalent bonds

4.3 STRUCTURE AND PROPERTIES

Plastics are polymerised organic substances . . . . . . of high molar mass . . . solids . . . . . . able to be shaped by flow.

When poly(ethene) is deformed, it stays in its new shape: it undergoes permanent or plastic deformation. Substances like this are called **plastics**. A plastic is a material of which the principal component is a polymerised organic substance of high molar mass, which is solid in its finished state and which, at some time in its manufacture or processing into finished articles, can be shaped by flow. Plastics find a large number of applications on account of their toughness, resistance to water and corrosive substances, ease of moulding and colour range. They are electrical and thermal insulators. Their melting temperatures are in general lower than those of metals and ceramics.

Elastomers return to their original shapes after being deformed, e.g. rubber. Fibres are polymers which can be made into strong thin threads, e.g. nylon.

Polymers which are soft and springy and return to their original shape after being deformed are called **elastomers**. Rubber is an example.

Strong polymers which do not change shape easily are used in the textile industry. They are made into thin, strong threads which can be woven together. These polymers are called **fibres**. Nylon is an example. Poly(propene) can be used as a plastic like poly(ethene) and it can also be made into a fibre for carpets.

4.3.1 THERMOSOFTENING AND THERMOSETTING PLASTICS

Since plastics soften on heating and harden on cooling, objects can be moulded easily from plastics. There are two subsets of plastics:

thermosoftening plastics (often called **thermoplastics**) and

thermosetting plastics (often called **thermosets**).

Thermosoftening plastics (thermoplastics) can be softened by heating, allowed to cool and harden, and then resoftened many times.

Thermosoftening plastics can be softened by heating, hardened and resoftened many times.

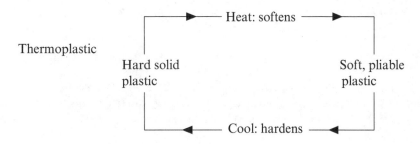

Thermoplastic

Heat: softens →

Hard solid plastic → Soft, pliable plastic

Cool: hardens ←

Thermosetting plastics are plastic during the first stages of manufacture, but once moulded they set and cannot be resoftened by reheating.

Thermosetting plastics, once set, cannot be softened by heating. In thermoplastics the forces of attraction between chains are weak.

Thermosetting plastic

Cool: hardens →

During manufacture, warm, pliable plastic → Permanently hard plastic

The reason for the difference in behaviour is a difference in structure. Thermoplastics consist of long polymer chains. The forces of attraction between chains are weak [see Figure 4.3A]. Thermosetting plastics have a different structure. When a thermosetting plastic is moulded, covalent bonds form between the chains. Cross-links are formed,

PTO

FIGURE 4.3A
The Structure of
(a) a Thermosoftening
Plastic,
(b) a Thermosetting
Plastic

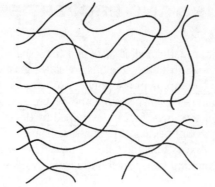 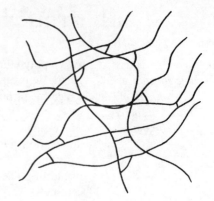

In thermosetting plastics, polymer chains are cross-linked to form a three-dimensional structure.

and a huge three-dimensional structure is built up. This is why thermosetting plastics can be formed only once.

Thermoplastics are very convenient for manufacturers to use. Manufacturers can buy tonnes of thermoplastic in the form of granules, soften the material, and mould it into the shape of the object they want to make. Moulding can be a continuous process [see §4.15]: as the plastic granules are fed into one end of the moulding machine, the moulded plastic comes out of the other end in the shape of tubes, sheets or rods. Plastics of this sort can be moulded several times during the manufacture of an article. Coloured plastics can be made easily by adding a pigment, melting the plastic and mixing thoroughly. Moulded objects made from the plastic are coloured all through. This is a big advantage over a coat of paint which can become chipped.

Thermoplastics are convenient to use in manufacturing as they are easily moulded.

The moulding of thermosetting plastics requires polymerisation to occur at the same time as moulding. The monomer is poured into the mould and heated to make it polymerise. As polymerisation occurs, a press forms the plastic into the required shape while it is setting. It is necessary to use a batch process. This is less efficient and more costly than a continuous process.

Thermosets are polymerised and moulded at the same time.

Both thermosoftening and thermosetting plastics can be strong, tough, rigid and stable towards chemical attack. Bonds within the molecule are covalent bonds. Between molecules and between segments of the same molecule intermolecular forces of attraction operate, e.g. van der Waals forces and forces of attraction between polar groups. When plastics melt or dissolve or flow, it is intermolecular bonds that are broken and made so that molecules and different segments of the same molecule can move past one another or away from one another. The strongest intermolecular bond, the hydrogen bond, operates in polyamides, e.g. nylon [see §4.12.3], and plays a big part in determining the properties of the polymer.

Within the polymer molecule covalent bonds operate. Between molecules and sections of the same molecule intermolecular forces of attraction operate. These may be van der Waals forces or electrostatic forces of attraction or hydrogen bonds.

Dipole–dipole interactions are present in e.g. polyesters:

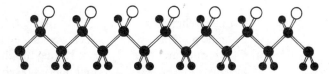

Polyester chain

Attraction between $\delta+$ of $C\overset{\delta+}{=}\overset{\delta-}{O}$ group and $\delta-$ of $C\overset{\delta+}{=}\overset{\delta-}{O}$ group

4.3.2 CONFIGURATION OF MOLECULES

Molecules may be linear or branched.

Individual molecules may be linear or branched. Linear molecules often take up configurations in which sections of the chain lie parallel to one another, giving the material regions of crystalline structure [see Figure 4.3B]. Branching interferes with the alignment of molecules and reduces crystallinity. It also reduces the ease of flow.

In 1954, Giulio Natta, an Italian chemist, used a Ziegler catalyst [see § 4.9.1] to polymerise propene. The product consisted of a crystalline poly(propene) and an amorphous form, which he was able to separate. In the crystalline form, the CH_3 groups all have the same orientation along the polymer chain. Natta named this the **isotactic form** (iso = same). In the amorphous polymer the methyl groups are randomly oriented, and he called this the **atactic form**. There are in fact three forms of poly(propene). In poly(propene), $(CH_2—CHCH_3)_n$, alternate carbon atoms in the chain are asymmetric. Since the arrangement of bonds about the carbon atom is tetrahedral, the possible structures (writing O for $—CH_3$) are

1. The isotactic structure

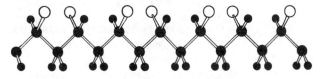

This regular structure has the CH_3 groups all on the same side of the carbon chain. The regularity of the structure leads to a high degree of crystallinity.

In polymers $(CH_2CHR)_n$, the arrangement of R groups may be isotactic or syndiotactic or atactic. These different structures give the polymers different physical properties.

2. The syndiotactic structure

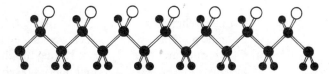

The alkyl groups alternate on each side of the chain in a regular manner in this structure.

3. The atactic structure PVC

In this structure the methyl groups have a random orientation.

Similarly, other polymers $(CH_2-CHR)_n$ exist as three structures. The way in which the R groups govern the stereochemical manner in which each new monomer adds to the growing chain is called **tacticity**.

Isotactic poly(propene) is crystalline and tough. It is used in sheets and films for packaging and as fibres in the manufacture of carpets. Atactic poly(propene) is soft and rubbery. It is used to make roofing materials, sealants and other weatherproof coatings.

Isotactic polymer molecules can be packed into a crystalline structure.

Many polymers are mixtures of **crystalline** (ordered) regions and **amorphous** (random) regions [see Figure 4.3B], in which the chains are further apart and have more freedom to move. A single polymer chain may have both crystalline and amorphous regions along its length.

FIGURE 4.3B
Crystalline and
Amorphous Regions
in a Polymer

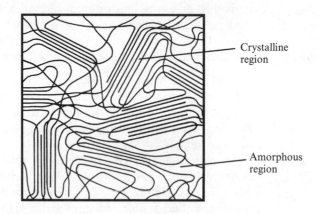

Crystalline region

Amorphous region

Many polymers have both crystalline and amorphous regions.
Linear-chain structures are more likely to have crystalline regions than branched chains or cross-linked chains.

Let us look at the ability of polymers to form crystalline regions.

● Linear-chain structures are most likely to form crystalline regions. Examples are isotactic poly(propene), in which polymer chains pack closely together because the CH_3 groups are regularly spaced along the chain, and Kevlar [§ 4.17.2], in which the absence of branched chains and bulky side chains make for crystallinity.

● Branched polymer chains are not easy to pack together in a regular manner, although if the branches are regularly spaced some crystallinity is possible. Poly(chloroethene) does not give rise to crystalline structures because the chlorine atoms are rather bulky and are irregularly spaced along the molecular chain.

● Cross-linked polymer chains cannot pack in a regular manner because of the links between the chains so crystallinity is not possible.

When an amorphous polymer is heated it does not show a definite melting temperature; it begins to soften. Since the arrangement of molecules in an amorphous polymer is disordered, as it is in a liquid, no structural change takes place when the polymer melts. For crystalline polymers there is an abrupt change at a certain temperature. Poly(ethene) has between 95% crystallinity (linear poly(ethene) with melting temperature 138 °C) and 60% (branched poly(ethene) with melting temperature 60 °C). Poly(chloroethene) with 0% crystallinity softens at 212 °C.

Liquid crystals are polymers.

Liquid crystalline polymers are described in *ALC*, § 6.9.

═══════════════════ **CHECKPOINT 4.3** ═══════════════════

1. (*a*) Explain the difference between a thermosoftening and a thermosetting plastic.

(*b*) What structural differences give rise to this difference in behaviour?

(*c*) Suggest two uses for which a thermosetting plastic would be the better choice.

(*d*) State two advantages to a manufacturer of working with a thermosoftening plastic.

2. (*a*) Write the formula of (i) poly(propene) and (ii) poly(propenoic acid).

(*b*) Draw structural formulae to show how the orientation of the carboxyl groups gives rise to three possible types of structure for poly(propenoic acid). Name each type of structure.

(*c*) What differences in properties would you expect between the three different types of structure?

3. Poly(propenol) was mentioned in §4.1.

(*a*) Write the formulae of propenol and poly(propenol).

(*b*) Show by means of a structural formula how water can bond to the polymer.

4.4 TENSILE STRENGTH

low *PVC*

The mechanical properties of a polymer are related to its degree of crystallinity. The more crystalline a polymer, the higher is its tensile strength [§1.4]. The relationship between the tensile strength of a polymer and the length of the chains is shown in Figure 4.4A.

FIGURE 4.4A
Tensile Strength as a Function of Chain Length

The tensile strength of a polymer depends on the degree of crystallinity . . .
. . . which increases with the length of the polymer chains . . .
. . . and decreases with chain-branching.

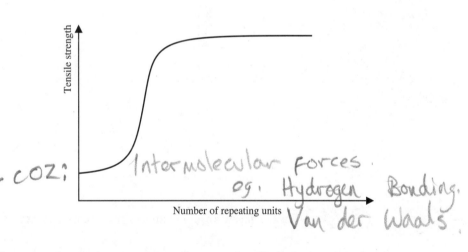

coz: Intermolecular forces. eg. Hydrogen Bonding. Van der Waals

The reason for the increase in tensile strength with chain length is that longer chains become more tangled together. The longer the chains are, the more points of contact they have with chains of neighbouring molecules, and the more intermolecular forces there are holding chains together. When a force is applied, no single chain can move without pulling other chains with it. This is why polymers have much greater tensile strengths than substances with smaller molecules of a similar kind. For hydrocarbon polymers, e.g. poly(ethene) the tensile strength increases when the chain length reaches about 100 repeating units. For nylon the tensile strength increases after about 40 units because the intermolecular bonds in nylon are much stronger.

Branching of the molecular chains also affects tensile strength. In the linear-chain form of poly(ethene) the molecules are densely packed to form a stiffer material with a higher tensile modulus. In the branched-chain form of poly(ethene), the molecules are packed less closely together and the polymer has a lower degree of crystallinity and a lower tensile modulus.

CHECKPOINT 4.4

1. (*a*) Explain what is meant by tensile strength.

(*b*) Explain why the tensile strength of a polymer does not increase steadily with chain length but only after a certain critical chain length is reached.

2. Explain why, after a certain length is reached, making

chains longer does not make much difference to the tensile strength of the polymer.

3. (*a*) What is the tensile modulus of a material?

(*b*) Why is it greater for linear-chain poly(ethene) than for branched-chain poly(ethene)?

4.5 GLASS TRANSITION TEMPERATURE

An amorphous polymer changes from a flexible material to a rigid material at the glass transition temperature.

Under normal conditions, polymers include soft, flexible substances and hard, brittle materials. If a soft, flexible polymer is cooled sufficiently a temperature is reached at which it becomes hard and glassy. A soft rubber ball when cooled in liquid air becomes hard and brittle and will shatter if it is dropped instead of bouncing. The temperature at which a polymer changes from a rigid to a flexible material is called the **glass transition temperature**. The material is considered to be changing from a glass-like material with a high tensile modulus to a rubber-like material with a low tensile modulus [see Figure 4.5A]. Below its glass transition temperature, the material has the type of stress–strain curve typical of a brittle substance; above the transition temperature the curve resembles that of a rubber-like material.

FIGURE 4.5A

A Plot of Stress against Strain for a Polymer Below and Above its Glass Transition Temperature, T_g

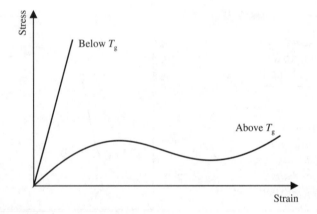

For the glass transition temperature of ceramics, see § 3.5.5.

4.5.1 GLASS TRANSITION TEMPERATURE AND USE

Amorphous polymers are shaped above the glass transition temperature and used below the glass transition temperature. Crystalline polymers are shaped above the melting temperature and used below the melting temperature.

Amorphous polymers are used below the glass transition temperature. They are formed and shaped at temperatures above the glass transition temperature when they are in a soft condition. Crystalline polymers are used above the glass transition temperature up to the melting temperature. They can be hot-formed and shaped at temperatures above the melting temperature or cold-formed between the glass transition temperature and the melting temperature [see Figure 4.5B].

High-density poly(ethene) is a crystalline polymer, melting temperature 138 °C and glass transition temperature −120 °C. It can be used up to 125 °C, just below the melting temperature. Poly(phenylethene), also called poly(styrene), is an amorphous polymer with a glass transition temperature of 100 °C. It is used up to 80 °C, just below the glass transition temperature. Other values of glass transition temperatures are: natural rubber −73 °C, low-density poly(ethene) −90 °C, poly(propene) −27 °C and poly(chloroethene) 87 °C.

4.5.2 GLASS TRANSITION TEMPERATURE AND STRUCTURE

The variation of tensile strength with temperature is important; for example a plastic spoon which is stiff at room temperature may not remain stiff when it is used to stir hot tea. The manner in which the tensile strengths of amorphous polymers change with temperature is the same for all such polymers, although the temperatures at which the changes occur are different [see Figure 4.5B].

FIGURE 4.5B
A Graph of Tensile Strength against Temperature for an Amorphous Polymer

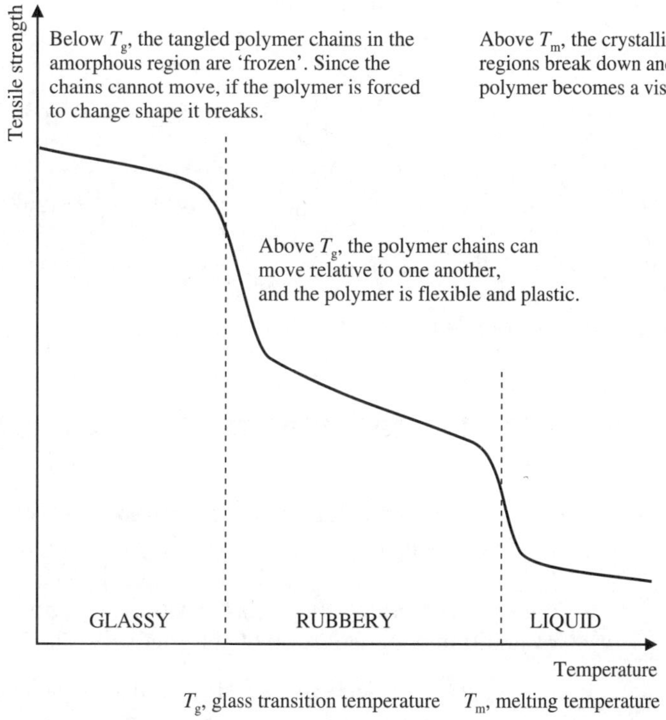

In a graph showing the decrease of tensile strength with temperature, three regions can be seen …
… corresponding to glassy, rubbery and liquid structures.

Below T_g, the tangled polymer chains in the amorphous region are 'frozen'. Since the chains cannot move, if the polymer is forced to change shape it breaks.

Above T_m, the crystalline regions break down and the polymer becomes a viscous fluid.

Above T_g, the polymer chains can move relative to one another, and the polymer is flexible and plastic.

GLASSY   RUBBERY   LIQUID

Temperature

T_g, glass transition temperature T_m, melting temperature

The type of molecular motion changes with decreasing temperature …
… from completely free rotation …
… to a regular crystalline structure …
… to an absence of rotation – a glass.

The reason for the change in tensile strength with temperature derives from the change in **molecular motion** with temperature. A polymer molecule is a long, regular chain in which the bonds are single bonds. At moderate temperatures, there is free rotation about each single bond, and the chain can take up a multitude of conformations in space. As the temperature falls, rotation becomes more restricted. When the temperature drops to the glass transition temperature, rotation about single bonds becomes impossible because of the energy barrier that a substituent group on one chain meets when it tries to move past a substituent group on an adjacent chain. The polymer molecules become entangled and trapped in a disordered state called a **glass**. The **glass transition temperature** is the temperature below which the free rotations cease; it is the temperature at which the polymer stiffens and changes from a flexible to a rigid material.

The glass transition temperature …
… at which a polymer changes from flexible to rigid …
… is higher for branched-chain polymers than for linear-chain polymers …
… and highest for cross-linked polymers.

CHAIN LENGTH AND BRANCHING

Polymers with linear-chain molecules tend to have lower glass transition temperatures than those with bulky side-chains, and these have lower glass transition temperatures than cross-linked polymers. The greater the degree of cross-linking, the higher the glass transition temperature. Thermosets often decompose before reaching the glass transition temperature.

Elastomers (rubbers; see §4.19) are lightly cross-linked polymers which under normal conditions are amorphous. They have glass transition temperatures below room temperature and are normally used above the glass transition temperature. If the temperature drops below the glass transition temperature, elastomers become stiff glassy solids of high tensile modulus.

Thermosets may decompose below the glass transition temperature. Elastomers are flexible only above the glass transition temperature.

In thermoplastics above the glass transition temperature the polymer chains in the amorphous regions of the material are held together by relatively weak van der Waals forces. The chains are able to slip into new permanent positions relative to one another when a stress is applied. The movement is not instantaneous; the speed at which it takes place depends on the viscosity of the material and is described as **viscoelastic deformation**.

CHECKPOINT 4.5

1. (*a*) What is the glass transition temperature of a plastic?

(*b*) How do the properties of a plastic differ above and below the glass transition temperature?

(*c*) Explain how this difference in properties arises from changes in the configuration of molecules.

(*d*) Why is it necessary to know the glass transition temperature before allocating a plastic to a particular use?

(*e*) Describe how you would determine the glass transition temperature of a plastic.

4.6. ADDITION POLYMERISATION

Polymerisation may take place by **addition** or by **condensation** [see § 4.12]. In **addition polymerisation**, many molecules of the monomer add to give one molecule of polymer. Poly(chloroethene), $(CH_2CHCl)_n$, is called PVC after its traditional name of poly(vinyl chloride). It is a plastic with many important uses [see § 4.7]. PVC is formed by addition polymerisation of chloroethene. The first step is:

$$2CH_2{=}CHCl \longrightarrow CH_3{-}CHCl{-}CH{=}CHCl$$

Polymerisation can take place by the addition of monomers containing C=C double bonds.

Polymerisation continues to give:

$$n\,CH_2{=}CHCl \longrightarrow (-CH_2{-}CHCl{-})_n$$

Addition polymerisation results from chain reactions. The steps involved are:

1. Generation of free radicals: Chain reactions require free radicals and are initiated by ultraviolet light and by the decomposition of compounds such as benzoyl peroxide.

$$C_6H_5C{-}CO{-}O{-}O{-}CO{-}C_6H_5 \longrightarrow 2C_6H_5COO\cdot \longrightarrow 2C_6H_5\cdot + 2CO_2$$

Benzoyl peroxide   Benzoyl radical   Phenyl radical

The decomposition can be initiated by heat or light. Thermal decomposition cannot be controlled very accurately because the system has thermal capacity and does not cool instantly when heating is stopped. Photochemical decomposition has the advantage that it can be controlled by altering the intensity of the initiating light to produce an immediate change in the rate of generation of free radicals.

2. Initiation: Free radicals $R\cdot$ add to $C{=}C$ bond.

The reaction starts with the generation of free radicals ...
... which initiate a chain reaction ...

$$CH_2{=}CHCl + R\cdot \rightarrow RCH_2{-}\overset{\displaystyle H}{\underset{\displaystyle Cl}{C}}\cdot$$

A new free radical is generated in the reaction.

3. Chain propagation: The free radical can add to another C=C bond.

$$CH_2{=}CHCl + RCH_2\overset{\bullet}{C}HCl \longrightarrow RCH_2CHClCH_2\overset{\bullet}{C}HCl$$

Chain propagation continues to form $R(CH_2CHCl)_n CH_2\overset{\bullet}{C}HCl$.

4. Termination: Chain propagation may continue until there is no monomer remaining. It is more usual, however, for the chain to be terminated before this by either (a) the combination of two radicals or (b) disproportionation.

(a)
$$RCH_2{-}\underset{\underset{Cl}{|}}{\overset{\overset{H}{|}}{C}}{\cdot} + {\cdot}\underset{\underset{Cl}{|}}{\overset{\overset{H}{|}}{C}}{-}CH_2R \rightarrow RCH_2{-}\underset{\underset{Cl}{|}}{\overset{\overset{H}{|}}{C}}{-}\underset{\underset{Cl}{|}}{\overset{\overset{H}{|}}{C}}{-}CH_2R$$

(b)
$$RCH_2{-}\underset{\underset{Cl}{|}}{\overset{\overset{H}{|}}{C}}{\cdot} + {\cdot}\underset{\underset{Cl}{|}}{\overset{\overset{H}{|}}{C}}{-}CH_2R \rightarrow RCH_2{-}\underset{\underset{Cl}{|}}{\overset{\overset{H}{|}}{C}}{-}H + \underset{\underset{Cl}{|}}{\overset{\overset{H}{|}}{C}}{=}CHR$$

... which is propagated ...
... or terminated by the addition or disproportionation of free radicals.

Other polymers formed by chain reactions include poly(propene), where CH_3 replaces Cl, and poly(phenylethene), polystyrene, where C_6H_5 replaces Cl.

4.6.1 CONTROL OF POLYMERISATION

The concentration of initiator determines the size of the polymer molecule.

Polymerisation is often first-order with respect to monomer and 1/2-order with respect to initiator. The concentration of the initiator is therefore important in determining the length of the polymer chain and must be carefully controlled.

As the rate of polymerisation increases the size of the polymer molecules increases.

As monomer is converted into polymer, the viscosity of the mixture increases. In a viscous medium, small monomer molecules are still able to diffuse and join growing polymer chains so the rate of propagation remains fairly constant. However, as the medium becomes more viscous, diffusion of the larger polymer molecules becomes more difficult. In consequence, the likelihood of large radicals meeting in termination reactions decreases, and the rate of the termination reaction decreases. The result is that the rate of polymerisation increases, and the size of the polymer molecules increases. The increase in rate with viscosity is an example of **autoacceleration**, in which the product of a reaction increases the rate of the reaction.

Polymerisation is exothermic so the reactor must be cooled.

Polymerisation is exothermic. The heat of reaction must be conducted away if polymerisation is to continue so the reactor is water-cooled. Most monomers and polymers are poor thermal conductors; therefore higher temperatures tend to reduce the relative molecular mass of the polymer. Should the temperature rise and a reactor vessel appear to be about to 'run away', 2-phenylpropene would be dumped into the reactor to mop up free radicals and terminate the reaction chain.

The manufacture of PVC has hazards. Chloroethene is hazardous to health. In the 1970s, workers exposed to chloroethene developed cancer of the liver. Subsequently working practices were changed to keep the level low (3 ppm of air), and safe ways of cleaning the reactor vessel were developed.

4.7 USES OF PVC

Poly(chloroethene), $(CH_2CHCl)_n$, PVC, is a more polar substance than poly(ethene). It is harder and stiffer than poly(ethene) because the chlorine atoms increase the forces of attraction between chains. PVC has a higher glass transition temperature than poly(ethene) and is a better electrical insulator. The structure is atactic, with only a small proportion of the polymer being crystalline.

Poly(chloroethene), PVC, is harder and stiffer than poly(ethene), with a higher glass transition temperature. Plasticisers are added to facilitate processing.

PVC is not easy to process when molten. The addition of **plasticisers** improves the properties and opens up a wider range of uses. A plasticiser is a substance with small molecules which can fit between polymer chains in such a way as to make them slide over one another more easily. The rigid unplasticised form, uPVC and the flexible plasticised form pPVC, have different uses, e.g. window-frames and cling-film [see Figures 4.7A and B]. Additives are chosen to improve strength and toughness.

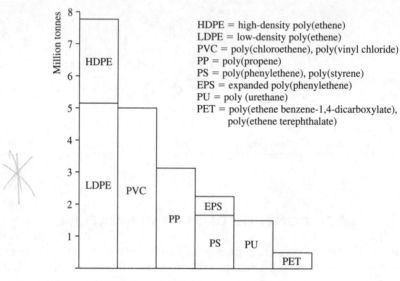

Each of the two forms of PVC – rigid unplasticised uPVC and flexible plasticised pPVC – has a wide range of uses . . . . . . which is extended by the incorporation of additives. Uses include building materials, water pipes, flooring, footwear, packaging for food etc., bottles and other containers.

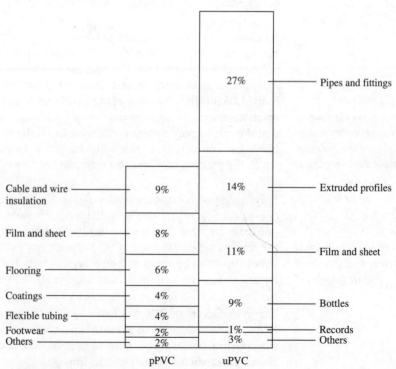

1. (*a*) Explain the terms (i) free radical, (ii) chain reaction, (iii) photochemical reaction [see *ALC*, §§ 14.7 and 26.3.8 if necessary].

(*b*) Explain the advantage of photochemical initiation over thermal initiation in a chain reaction.

2. Ethene polymerises by a chain reaction which involves free radicals.

(*a*) Write the formula of the free radical that is formed when the free radical R • adds to a molecule of ethene.

(*b*) Explain how this free radical can give rise to a chain reaction.

(*c*) Why does the chain terminate before all the ethene has reacted?

3. Compare the physical properties of poly(ethene) and poly(chloroethene). How can the difference be explained on structural grounds?

4. Explain why poly(chloroethene), PVC, is a versatile plastic with many different uses. Mention six uses of PVC and explain how its properties fit it for these uses.

4.8 LOW-DENSITY POLY(ETHENE)

4.8.1 MANUFACTURE OF LDPE

Low-density poly(ethene) ldpe, is made at high pressure ... ... has branched molecules ... ... a low degree of crystallinity, ... lower values than hdpe of ... ... density stiffness, tear strength, hardness, yield stress, softening temperature.

Low-density poly(ethene), ldpe, with density 0.91–0.94 g cm^{-3}, is made at high pressure (about 15 atm). It has branched molecules with both short-chain branching and long-chain branching. The properties of a particular poly(ethene) depend on the molar mass, the length of the chains and the extent of branching. The density of a polymer depends on the degree of crystallinity. Short-chain branching has these effects:

- a decrease in the extent of crystalline regions,

- and therefore a decrease in density,

- and a decrease in other properties which depend on crystallinity: stiffness, tear strength, hardness, resistance to chemical attack, softening temperature and yield stress,

- and an increase in toughness and permeability to liquids and gases.

4.8.2 USES OF LDPE

The uses of ldpe depend on its toughness, durability and resistance to water, acids and alkalis. Uses include plastic bags, cable insulation, toys and housewares.

Uses of low-density poly(ethene), ldpe, are many because the material is tough and durable and resists attack by water, weather, acids and alkalis. It is used in films to make plastic bags and to wrap produce, frozen foods and items of clothing. Thicker sheets are used for pond liners, tank liners and waterproof covers. Electrical cables are insulated with poly(ethene) because it withstands bad weather conditions better than rubber which was used previously for this purpose. One of the first applications of poly(ethene) was as an insulation for a telephone cable between the Isle of Wight and the mainland.

4.9 HIGH-DENSITY POLY(ETHENE), HDPE

4.9.1 MANUFACTURE OF HDPE

The manufacture of poly(ethene) described in § 4.8.1 requires the construction of a costly plant which will withstand high pressure. The situation changed in 1953 with a major discovery made by Karl Ziegler, a German chemist. He found that, at

High-density poly(ethene), hdpe, is made at a pressure slightly above 1 atm, by polymerisation in a solvent ...
... in the presence of a Ziegler catalyst.

atmospheric pressure, ethene passed into a dilute solution of titanium(IV) chloride and triethylaluminium (or a similar alkylaluminium compound) in a liquid alkane polymerised immediately. The catalysts he used are named **Ziegler catalysts**. The poly(ethene) produced was different from that obtained at high pressure: it had a higher density, 0.95–0.97 g cm^{-3}.

High-density poly(ethene) is manufactured at a pressure slightly above atmospheric pressure at 50–75 °C in a solvent such as heptane in the presence of a Ziegler catalyst, e.g. titanium(IV) chloride and triethylaluminium. The heat liberated during the reaction is removed by cooling. The polymer forms as an insoluble powder or as granules. When polymerisation is complete, the catalyst is hydrolysed and the polymer is filtered off, washed and dried.

Hdpe has linear molecules with little branching ...
... which pack into a largely crystalline structure ...
... with a high density.

High-density poly(ethene), hdpe, is composed of long molecules with very little branching: less than one side-chain per 200 carbon atoms in the main chain. The molar mass is about 3×10^6 g mol^{-1}. The chains can pack closely together into a largely crystalline structure, giving the polymer a higher density. Compared with low-density poly(ethene), high-density poly(ethene) is harder and stiffer, with a higher melting temperature (about 135 °C) and greater tensile strength. It has good resistance to chemical attack, is brittle at low temperature and has low permeability to gases.

FIGURE 4.9A
The Structure of Low-Density Poly(ethene)

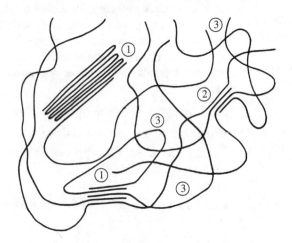

1 Crystalline region with folding
2 Crystalline region without folding
3 Amorphous region

FIGURE 4.9B
The Structure of High-Density Poly(ethene)

Hdpe is stiffer, harder, stronger than ldpe ...
... has a higher melting temperature ...
... and is resistant to chemical attack.

The chains are aligned and packed closely together.

4.9.2 USES OF HDPE

High-density poly(ethene), hdpe, is less easily softened by heating than ldpe. Hdpe is used in the production of bottles and other containers by blow-moulding and in the production of water tanks and pipes, crates, tubs, washing-up bowls, food containers and other housewares by injection moulding. Hdpe is strong and easily moulded into complicated shapes and is even used for vehicle petrol tanks. Since it can be heat-sterilised, hdpe is used in the manufacture of hospital equipment such as trays, buckets and bed pans.

FIGURE 4.9C
Disposable Hospital Ware

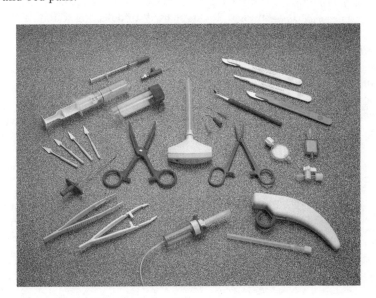

Uses of hdpe include . . .
. . . water tanks and pipes . . .
. . . bottles and other containers, housewares,
. . . hospital equipment which can be sterilised.

4.10 POLY(PHENYLETHENE), POLY(STYRENE)

Styrene is the traditional name for phenylethene, $C_6H_5CH{=}CH_2$. It is made from benzene and ethene, which react to form ethylbenzene.

Phenylethene is made from benzene and ethene . . .
. . . and polymerised to poly(phenylethene) . . .
. . . known as poly(styrene).

$$\langle\bigcirc\rangle + CH_2{=}CH_2 \xrightarrow{\text{AlCl}_3,\ 95\ °C} \langle\bigcirc\rangle{-}CH_2CH_3$$

The next step is catalytic dehydrogenation:

$$\langle\bigcirc\rangle{-}CH_2CH_3 \xrightarrow{\text{ZnO + steam, }630\ °C} \langle\bigcirc\rangle{-}CH{=}CH_2 + H_2$$

Finally phenylethene is polymerised to form poly(phenylethene) [see § 4.13].

Uses of poly(phenylethene) include . . .
. . . disposable cups and cartons . . .
. . . plastic foam used in packaging.

Poly(phenylethene) softens at 100 °C and is one of the easiest plastics to mould. It degrades above 150 °C. Since it is transparent, it is usual to add a pigment to give an opaque material. Crystalline poly(styrene) is familiar as ball-point pens. Disposable coffee cups and cartons for dairy produce are made of polystyrene with the white pigment titanium(IV) oxide. Poly(styrene) foam is made by adding hexane to the solid beads of polymer before they are heated and moulded. It is widely used for packaging breakable objects.

4.11 OTHER ADDITION POLYMERS

Poly(tetrafluoroethene), PTFE, is important for its non-stick properties and as insulation. Poly(1,1-difluoroethene) is piezoelectric. Poly(methyl 2-methylpropenoate), Perspex, is transparent. Some plastics can be 'doped' to make them conduct electricity.

Poly(tetrafluoroethene), PTFE, is an addition polymer of tetrafluoroethene, $CF_2{=}CF_2$. PTFE is thermoplastic, chemically unreactive and non-flammable. It has very low friction, and its non-stick properties find it uses as a coating for skis, saucepans, etc. It is used as cable insulation, as sealing tapes for plumbing joints and for making valves and bearings.

Poly(1,1-difluoroethene) is **piezoelectric**; that is, it generates electricity when bent or twisted. It can be used for detecting sound. If a sound wave impinges on a sheet of poly(1,1-difluoroethene) the sheet vibrates, generating electric signals which can be transmitted to a detector.

Perspex is an addition polymer of methyl 2-methylpropenoate (formerly called methyl methacrylate), $CH_2{=}C(CH_3)CO_2CH_3$. It is an important plastic because it is transparent and can be used instead of glass. It is more easily moulded than glass and less easily shattered. It is used for motor cycle windscreens and safety glasses.

4.11.1 CONDUCTING PLASTICS

When a stream of ethyne is directed on to the surface of a Ziegler catalyst [§4.9.1] at $-78\,^{\circ}\mathrm{C}$, *cis*-poly(ethyne), $\overline{(\text{CH}{=}\text{CH})_n}$, is formed.

Cis-poly(ethyne) is red, unstable and on warming is converted into the *trans*-form.

Trans-poly(ethyne) is blue and is the stable form of the polymer.

Some polymers can be 'doped' to make electrical conductors . . .

The alternating double and single bonds mean that π-bonds are formed between adjacent carbon atoms, and a band of delocalised electrons stretches the full length of the molecule. Such a chain is an insulator. To make it into a conductor, some electrons must be removed from the band of delocalised electrons. This is done by doping – adding an oxidising agent, e.g. iodine, which will remove some electrons.

. . . e.g. poly(ethyne) . . .

On doping, the conductivity increases to a value of $1 \times 10^4\ \mathrm{S\,cm^{-1}}$, which is not far below that of a metal.

In pyrrole, C_4H_5N, two electrons from each double bond and a lone pair of electrons on the nitrogen atom are delocalised over the ring. In poly(pyrrole) delocalisation extends along the whole length of the polymer chain.

. . . e.g. poly(pyrrole)

Pyrrole   Poly(pyrrole)

Poly(pyrrole) can be made into a conductor by doping with e.g. iodine.

Conducting plastics are sometimes called 'synthetic metals'. They combine the conductivity of metals with the properties of plastics: flexibility, low density, freedom from corrosion and low cost. They could lead to plastic batteries and use in computer displays. Poly(ethyne) is an inconvenient material for fabrication into commercial electronic devices. It cannot be melted and shaped into required articles, and it is oxidised on contact with air. Many other doped conjugated polymers are conducting, and research workers are looking for better mechanical properties. A plastic battery can be made with polymer sheets acting as both anode and cathode. Polymer electrodes should last longer than metal electrodes because the ions which deliver and store charge are part of the electrodes themselves. The mechanical wear which results from dissolution and redeposition of metal in traditional batteries does not occur.

CHECKPOINT 4.11

1. (a) Outline the difference in manufacture of low-density poly(ethene), ldpe, and high-density poly(ethene), hdpe.

(b) Describe the difference in structure between the two polymers and explain how this explains the difference in properties.

(c) Give three examples of the uses of ldpe and three for hdpe.

2. Some properties of low-density poly(ethene), ldpe, and high-density poly(ethene), hdpe, are shown in the table.

Plastic	Tensile strength/ Mpa	Elongation at fracture/%	Price (1994)/ (£ kg^{-1})
ldpe	15	600	0.70
hdpe	29	350	0.85

(a) Why is the tensile strength of hdpe greater than that of ldpe?

(b) Why is hdpe more expensive than ldpe?

(c) How does the difference in physical properties shown above explain the difference in the uses of ldpe and hdpe?

3. (a) Write the structural formula of

(i) poly(phenylethene) – poly(styrene),

(ii) poly(tetrafluoroethene) – PTFE

(iii) poly(methyl 2-methylpropenoate) – Perspex.

(b) Give two examples of the uses of each polymer. State what advantage the plastic has over the material that was used for the purpose before plastics came on the scene.

4.12 CONDENSATION POLYMERISATION

Condensation polymerisation gives a polymer of formula $(P)_n$ which is not the same as n times the formula of the monomer, M. The molecule P_n lacks some of the atoms that are present in n molecules of M. When molecules of monomer combine, molecules of water or another compound are eliminated. Condensation polymers are formed by an addition–elimination reaction between reactive groups.

4.12.1 POLYESTERS

A **polyester** is formed by condensation – the elimination of water – between a diol and a dicarboxylic acid; that is, between two bifunctional compounds.

A dicarboxylic acid and a diol react to form a diester.

Condensation polymerisation involves combination of molecules of monomer to form a large molecule of polymer with the elimination of molecules of water or another compound.

Benzene-1,4-dicarboxylic acid   +   Ethane-1,2-diol

Further condensation gives the polyester, $\left[\!\!\!-CO-\!\!\bigcirc\!\!-CO_2CH_2CH_2O-\!\!\!\right]_n$

Polyesters are formed by condensation polymerisation between a diol and a dicarboxylic acid, e.g. Terylene or Dacron.

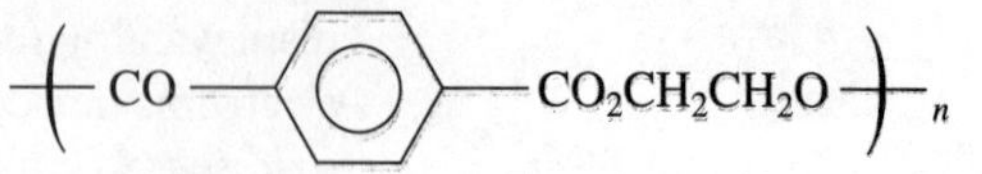

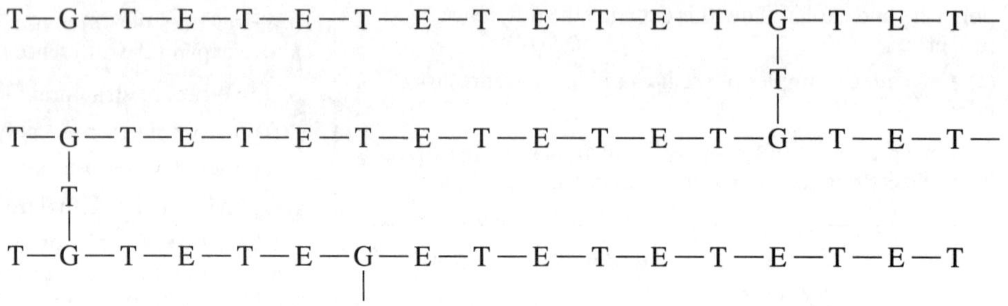

The product of the reaction shown above is the polyester with the ICI trade name Terylene and the DuPont trade name Dacron. The name comes from *ter*ephthalic acid (the traditional name for benzene-1,4-dicarboxylic acid) and ethyl*ene* glycol (the traditional name for ethane-1,2-diol). Since the carboxyl groups are in the 1,4-positions, the product is a linear molecule, ideal for the formation of a fibre. If some propane-1,2,3-triol (glycerol) is included in the reaction mixture, a cross-linked product is formed. Writing T for benzene-1,4-dicarboxylic acid, E for ethane-1,2-diol and G for propane-1,2,3-triol, the structure can be shown as:

A tougher polyester is formed when some triol is included in the reaction mixture to form cross-links.

```
T—G—T—E—T—E—T—E—T—E—T—E—T—G—T—E—T—
            |
            T
            |
T—G—T—E—T—E—T—E—T—E—T—E—T—G—T—E—T—
    |
    T
    |
T—G—T—E—T—E—G—E—T—E—T—E—T—E—T—E—T
    |
```

Cross-linked polymers are harder and tougher than straight-chain polymers. Their properties can be varied by regulating the amount of cross-linking. **Polycarbonates** are a subset of polyesters with the general formula

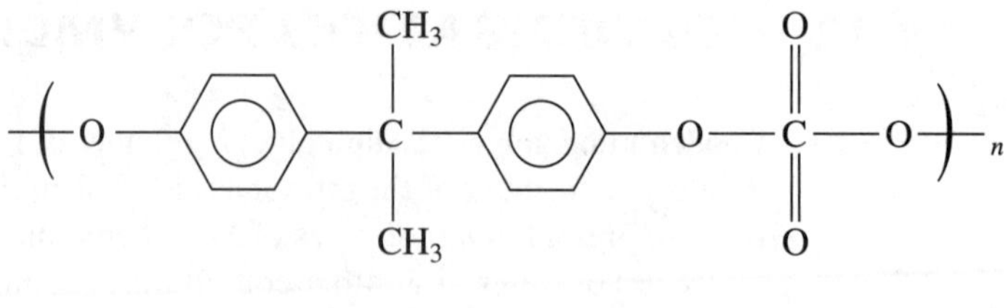

$$\left[\!\!\!-O-\!\!\bigcirc\!\!-\overset{\overset{\displaystyle CH_3}{|}}{\underset{\underset{\displaystyle CH_3}{|}}{C}}-\!\!\bigcirc\!\!-O-\overset{\overset{\displaystyle O}{\|}}{\underset{\underset{\displaystyle O}{\|}}{C}}-O-\!\!\!\right]_n$$

4.12.2 POLYESTER RESINS

Polyester resins are formed by a condensation reaction between a diol and a dicarboxylic acid or anhydride. If one of the reactants is unsaturated, an unsaturated polymer is formed.

Polyester resins are formed when either the diol or the dicarboxylic acid is unsaturated.

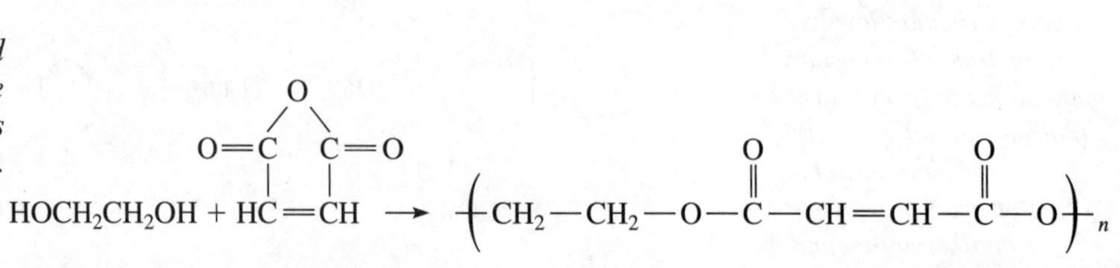

$$HOCH_2CH_2OH + \overset{\overset{\displaystyle O}{/\,\backslash}}{\underset{\underset{\displaystyle }{|\quad|}}{O=C\quad C=O}}\quad \longrightarrow \left[\!\!\!-CH_2-CH_2-O-\overset{\overset{\displaystyle O}{\|}}{C}-CH=CH-\overset{\overset{\displaystyle O}{\|}}{C}-O-\!\!\!\right]_n$$

Ethane-1,2-diol   Butane-1,4-dioic anhydride   Unsaturated polyester

An unsaturated polyester can form cross-links. For example, X—CH=CH—Y can add to the polyester shown above to form a **thermosetting polyester resin** with the cross-linked structure:

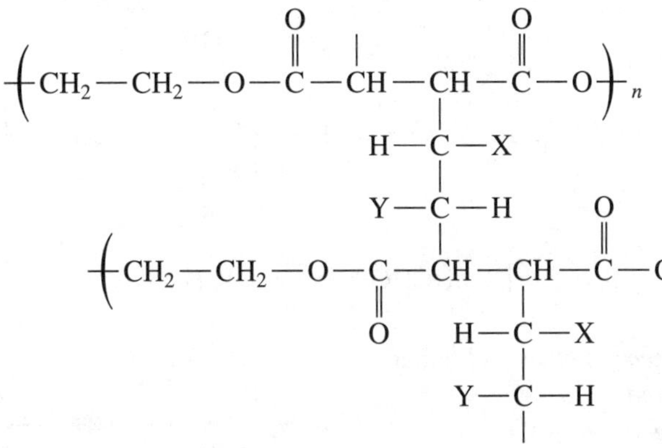

The formation of cross-links gives rise to a thermosetting polyester resin.

4.12.3 POLYAMIDES: NYLONS

A dicarboxylic acid and a diamine react to form a **polyamide**. The reaction between hexanedioic acid and hexane-1,6-diamine begins with the step:

Polyamides, nylons, are formed by condensation polymerisation between a dicarboxylic acid and a diamine.

Nylon 6,6 is made from hexane-1,6-diamine and hexanedioic acid.

$$H_2N(CH_2)_6NH_2 + HO_2C(CH_2)_4CO_2H \longrightarrow H_2N(CH_2)_6NHCO(CH_2)_4CO_2H + H_2O$$

Hexane-1,6-diamine   Hexanedioic acid

The product, having an amino group and a carboxyl group, can polymerise further to form $H_2N(CH_2)_6[NHCO(CH_2)_4CONH(CH_2)_6]_nNHCO(CH_2)_4CO_2H$. This is **nylon 6,6**. The name comes from the numbers of carbon atoms in the monomers. Nylon 6,10 is made from hexane-1,6-diamine and decanedioic acid. Nylon 6 is $—[(CH_2)_5CONH]_n—$.

4.12.4 PHENOLIC RESINS

Another condensation reaction is that which occurs between phenol and methanal to form a phenol-methanal resin. The first step can be shown as:

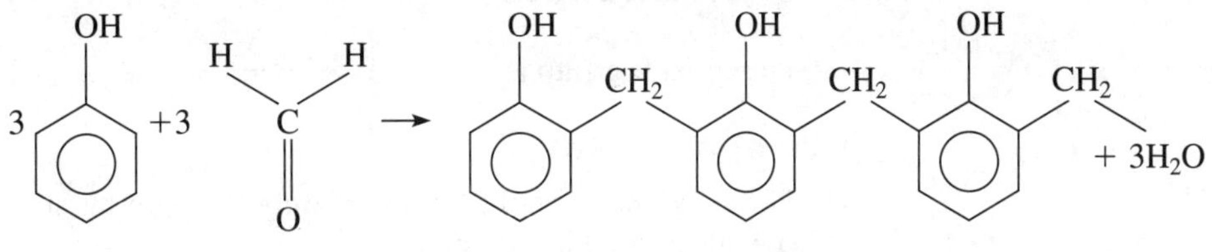

Phenol   Methanal

If the proportion of phenol in the reaction mixture is reduced, the 4-position is also utilised with the formation of the polymer

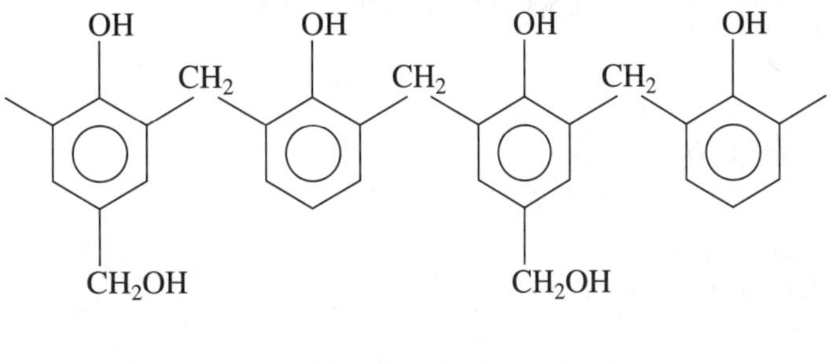

On further heating, cross-linking takes place with the formation of the phenol–methanal resin.

A phenolic resin is formed by condensation polymerisation between phenol and methanal to form a cross-linked thermosetting plastic, e.g. Bakelite.

A phenol–methanal resin (formerly called phenol–formaldehyde resin) was discovered by Leo Baekland, a Belgian-born American in 1872. He described it as 'perfectly insoluble, infusible, unaffected by almost all chemicals, an excellent insulator for heat and electricity with density $1.25\,g\,cm^{-3}$. Baekland patented his discovery, which he named Bakelite, and set up a company to manufacture it. Phenol–methanal resins are thermosetting polymers. The uses of these resins include radio cases, electrical switches and sockets, car dashboards, saucepan handles, fuse holders and door handles. Bakelite is easily recognisable by its brown colour. It has to a large extent been replaced by urea–methanal resins.

Phenol–methanal resins are used in the production of laminates.

Phenol–methanal resins are used in the production of **laminates** (layered materials; see § 4.12.6). [For laminates see also § 5.9.]

4.12.5 EPOXY-RESINS

The epoxide group is $-CH-CH-$
                              $\diagdown O \diagup$

The angles between the bonds are only $60\,°$ and this strained configuration makes the epoxide group highly reactive.

The epoxide of 3-chloropropane $CH_2-CH-CH_2Cl$
                                        $\diagdown O \diagup$

Epoxy resins are made by condensation polymerisation of a compound with an epoxide group, e.g. $CH_2-CH-CH_2Cl$,
$\diagdown O \diagup$
with a compound with two phenolic groups.

condenses with compounds with two phenolic groups, e.g.

The elimination of hydrogen chloride gives a polymer with the repeating unit

$$\left(\!\!O\!-\!\!\bigcirc\!\!-\!\!\underset{\underset{CH_3}{|}}{\overset{\overset{CH_3}{|}}{C}}\!-\!\!\bigcirc\!\!-\!\!O\!-\!CH_2\!-\!\underset{\underset{\;}{|}}{\overset{\overset{OH}{|}}{CH}}\!-\!CH_2\!\right)_{\!n}$$

Epoxy resins are hard tough solids used as adhesives, electrical insulators, etc.

If n is 25 or more, the resin is a hard tough solid. Any resin containing one or more epoxide groups is an **epoxy resin**. Epoxy resins are intermediates which must be **cured**, that is, **cross-linked**, to yield useful resins. Epoxy resins have uses which range from adhesives, e.g. Araldite, to can and drum coatings. Strong and tough, with excellent resistance to chemicals, they are used as floorings, linings, surface coatings and in laminates [see § 5.9].

4.12.6 FORMICA AND MELAMINE

Urea, $CO(NH_2)_2$ and methanal, HCHO, condense to form a polymer, **urea–methanal**:

Urea and methanal condense to form a polymer, a hard thermosetting

$$n\text{HCHO} + n\text{H}_2\text{NCONH}_2 \longrightarrow -(\text{CH}_2\text{NHCONH})_n - + n\text{H}_2\text{O}$$

Methanal   Urea   Urea–methanal polymer

Cross-linking produces a thermosetting resin which is used for trays, knobs, toilet seats, etc. and also in laminates. It is often known by its older name of urea–formaldehyde resin and by its trade name of **Formica**.

Melamine

Urea and melamine condense to form the cross-linked thermosetting plastic used to make tableware under the names of Melamine® and Melaware®

Methanal also condenses with Melamine to form a highly cross-linked polymer resin, **methanal–melamine**. It is thermosetting and, with its qualities of hardness, resistance to chemical attack and good finish, is used for tableware with the trade name Melamine. It is also used in laminates.

A **laminated material** is formed from layers of material bonded together. Laminated plastics are made from thermosets, e.g. Melamine, epoxy resins, unsaturated polyester resins and phenolic resins. The resin is dissolved in a suitable solvent, and the filler is coated or impregnated with the same solvent. Curing is effected by compressing the resin and filler in a mould at 7–17 MPa and 130–170 °C.

Many laminated materials are made by impregnating layers of materials with thermosetting plastics and curing.

Many different fillers are used. They include paper pulp and wood 'flour', shredded textiles, synthetic aramid fibres, calcium carbonate, silica, metal oxides and silicates and filaments such as fibreglass and aluminium oxide. [For laminates see also § 5.9.]

4.12.7 POLYSULPHONES

Polysulphones contain the unit

$$R\!-\!\overset{\overset{O}{\|}}{\underset{\underset{O}{\|}}{S}}\!-\!R'$$

Polysulphones are useful at high temperatures.

where the groups R and R′ can be aliphatic or aromatic. Their stability at high temperatures enables them to be sterilised in autoclaves, and they can be used for medical equipment. They are also used for microwave cookware and corrosion-resistant piping.

CHECKPOINT 4.12

1. (*a*) Give an example of a pair of compounds which can react to form a polyester.

(*b*) What structural feature is needed for the formation of a cross-linked polyester?

(*c*) What is the difference in properties between a cross-linked polyester and a straight-chain polyester?

2. (*a*) Give an example of a pair of compounds which react to form a polyamide.

(*b*) By what other name are polyamides known?

3. Phenol–methanal resins are thermosetting polymers.

(*a*) What structural feature is responsible for the thermosetting property?

(*b*) Briefly describe how a phenol–methanal resin is used to make a laminated material.

4. (*a*) Name and give the formula of a compound which can condense with a phenol to form an epoxy resin.

(*b*) Give two applications of epoxy resins.

4.13 TECHNIQUES OF POLYMERISATION

Polymerisation may be carried out by a batch process or by a continuous process. Both methods have advantages.

Most polymerisations are carried out in the liquid phase. A **batch process** or a **continuous process** may be used. A continuous method has the advantages of giving a more uniform product and having lower operating costs. On the other hand, the difficulties in operating a continuous method are:

● Some catalyst may remain in the polymer and degrade it during processing.

● The polymer may have a wide distribution of molar mass.

● Polymer may stick to the reactor walls and necessitate a shutdown for cleaning.

Five methods of polymerisation are in use.

1. MASS POLYMERISATION OR BULK POLYMERISATION

In mass polymerisation or bulk polymerisation, no solvent is used; only a catalyst and the monomer ...
... which may be gaseous. A batch process may be used ...
... or a continuous operation, may be used.

No solvent or diluent is used – only the monomer and a catalyst. A Ziegler catalyst is used [see § 4.9.1]. An initiator is added to a mass of the monomer in the liquid or vapour state in a reaction vessel fitted with a stirrer. Polymerisation is exothermic and provision must be made to remove the heat generated. An advantage of bulk polymerisation is that there is no solvent to be recovered and purified. A difficulty is that if the polymer dissolves in the liquid monomer, the viscosity increases as polymerisation proceeds and it becomes more difficult to conduct away the heat evolved. In addition, autoacceleration occurs [see § 4.6.1] because the high viscosity reduces the rate of collisions between free radicals which would bring about chain termination. If the polymer is insoluble in the monomer a slurry is formed. A continuous process is employed, with the reaction mixture passing through a series of reactors at increasing temperatures. Polymerisation rarely reaches 100% conversion. Unconverted monomer is removed by distillation under reduced pressure and recycled. Mass polymerisation gives polymers of high molar mass. It is used for condensation polymers, e.g. nylons, and also for poly(propenenitrile), poly(phenylethene), nylon, PVC, poly(ethene) and poly(propene).

In solution polymerisation ...
... a batch process ...
... the monomer and the polymer both dissolve in the solvent ...
... from which the product is obtained by distillation.

2. SOLUTION POLYMERISATION

A solvent is often used when the heat evolved in bulk polymerisation is too great to be controlled. Temperature control is easier because the solvent adds to the heat capacity. Autoacceleration is less prevalent because the solvent reduces viscosity and free radicals are more able to collide and combine. The solvent and the monomer are removed from the polymer by distillation. Solution polymerisation gives polymers of low to medium molar mass. This method is used for addition polymers, e.g. poly(propenenitrile), poly(methyl 2-methylpropenoate), poly(phenylethene) and also for nylon. It is usually operated as a batch process.

3. SLURRY POLYMERISATION

In slurry polymerisation, the monomer and the initiator dissolve in the solvent but the polymer is insoluble and precipitates as it is formed. The process can run continuously.

The solvent which is used dissolves the monomer and the initiator but not the polymer. As polymerisation starts, polymer particles start to precipitate from solution. No stabiliser or suspending agent is required (as in suspension and emulsion polymerisation) and the polymer is pure. Ethene and propene are polymerised by this method.

4. SUSPENSION POLYMERISATION

In suspension polymerisation, droplets of monomer are suspended in water.
As polymerisation takes place, each droplet forms a bead of polymer.
It is run as a batch process.

If the monomer is insoluble in water, polymerisation can be carried out in suspended droplets. The droplets are prevented from coagulating by adding a suspending agent, e.g. poly(ethenol) and by vigorous stirring. Small spherical beads (about 1 mm diameter) of polymer separate out when stirring ceases. The reaction is run as a batch process. Suspension polymerisation gives polymers of relatively low molar mass. Examples are PVC and poly(phenylethene).

5. EMULSION POLYMERISATION

Emulsion polymerisation is used for polymers of high molar mass . . .
. . . and run as a batch process.

The liquid monomer forms an emulsion in water, and the polymer is insoluble in water. Water-soluble initiators are used. The particles are much smaller in emulsions ($0.05\,\mu m$–$5\,\mu m$ diameter) than they are in suspensions ($10\,\mu m$–$1000\,\mu m$ diameter). An emulsifier is added, and vigorous stirring is employed. The product cannot be filtered off; it is an emulsion of small colloidal particles (about $0.1\,\mu m$ diameter) of polymer in water – a latex. Emulsion polymerisation gives polymers of very high molar mass. Examples are PVC, poly(ethenyl ethanoate) (PVA), poly(phenylethene) and its copolymers. A batch process is employed.

4.13.1 CONTROL OF POLYMERISATION

If there is a relatively high proportion of initiator, the polymer chains will terminate fairly soon and be relatively short. If there is a relatively low proportion of initiator, polymers of long chain length will be formed. The amount of initiator is therefore controlled to give a polymer with the required chain length.

The amount of initiator in the reaction mixture is important. A high proportion of initiator leads to short polymer chains.

The temperature rises as polymerisation proceeds. Unless the vessel is cooled, the rate of reaction will increase and it will become difficult to control conditions that give the required chain length. Initiators which decompose readily, such as benzoyl peroxide, will be used up early in the reaction. Additional initiators which decompose at higher temperatures are added to take over when the less stable initiators have been used up.

4.13.2 POLY(PHENYLETHENE)

The three processes used in the manufacture of poly(phenylethene) are:

- mass polymerisation: The monomer is removed and the product is distilled.
- solution polymerisation: The solvent is removed, and the product is distilled.
- suspension polymerisation: Water is removed, and the product is dried.

The separations consume energy and are costly. In addition, a powerful stirrer must be used because the mixture becomes viscous as reaction proceeds. The reaction vessel has to be loaded, heated and later cooled for each batch of product. Then it is unloaded and cleaned before the next batch. A continuous process would save energy and labour and would increase capacity.

A continuous method of mass polymerisation has been developed for poly(phenylethene).

British Petroleum has developed mass polymerisation of phenylethene into a continuous process.. The monomer is fed into a reactor at 140 °C where a conversion of 45% occurs. The mixture passes to a second reactor at 160 °C, where 70% conversion is reached. The second reactor needs a powerful stirrer to cope with the viscous mixture. The mixture is then distilled under reduced pressure to remove the monomer for recycling and yield the polymer.

4.14 METHODS OF MODIFYING THE PROPERTIES OF PLASTICS

Methods of modifying the properties of plastics include . . .

Metals are changed by heat treatment [see § 2.11]. For polymers the methods employed to change properties are different. They include the following:

1. Increasing the length of the molecular chain or introducing large side-groups or producing chain-branching. All these changes increase tensile strength and tensile modulus because the chains can move less easily.

. . . modifying the molecular structure . . .

2. Cross-linking chains. The greater the degree of cross-linking, the more chain motion is inhibited and the more rigid is the material.

3. Making some of the material crystallise, e.g. by stretching to align polymer molecules in a particular direction. The greater the degree of crystallinity, the denser the material and the higher its tensile strength and tensile modulus.

. . . addition of plasticisers etc. . . .

4. The addition of plasticisers, liquids which fill some of the spaces between polymer chains, makes it easier for the chains to move and increases flexibility.

. . . copolymerisation . . .

5. Copolymerisation combines two or more monomers to make a single polymer with different properties. For example,

● If an application requires poly(propene) with a slightly lower glass transition temperature, a suitable copolymer can be made by adding ethene to propene during polymerisation.

● Poly(phenylethene) can be copolymerised with propenenitrile and buta-1,3-diene to give ABS, a tough, solvent-resistant plastic used for telephone casings etc.

● Poly(chloroethene), PVC, can be improved for such purposes as making drainpipes by lowering its melting temperature. This can be done by copolymerising with ethenyl ethanoate $CH_3CO_2CH=CH_2$. Larger amounts of this copolymer give a more flexible PVC which is used for floor tiles etc.

. . . blending to form polymer alloys . . .

6. Polymer alloys can be made by mixing molten polymers and allowing them to solidify together. The mixture is not homogeneous. In contrast to metals, homogeneous mixtures of polymers are rare.

. . . and injecting gases to form plastic foams.

7. Plastic foams: Sometimes gases are mixed with softened plastics to make low-density plastic foams. These materials are used in car seats, for thermal insulation of buildings, for insulation against sound and for packaging. For some years the chemically unreactive gases, chlorofluorohydrocarbons, CFCs, were used in plastic foams. Then the danger to the ozone layer from CFCs was established [see *ALC*, § 29.9]. Now hexane is used for this purpose. Since hexane is flammable, care must be taken to avoid setting fire to plastic foams.

4.15 METHODS OF MOULDING PLASTICS

Different methods are used for moulding thermosoftening and thermosetting plastics [see § 4.3]. Five of the methods of moulding thermosoftening plastics are shown in Figures 4.15A–F. They are **injection moulding, extrusion moulding, blow moulding, extrusion stretch blow moulding** and **vacuum forming. Pressure forming** is the opposite of vacuum forming in that the plastic is blown into shape by compressed air instead of being sucked into shape by reduced pressure. Large sheets of plastic are made by **calendering**: passing softened thermoplastic over heated pressure rollers. Car seat covers and floor coverings are made in this way. The **compression moulding** method used for thermosetting plastics is shown in Figure 4.15G and the method of making laminated thermosetting resins in Figure 4.15H. In moulding plastics, there is a danger of thermal degradation. The methods used avoid high temperatures and use fast processing.

FIGURE 4.15A
Injection Moulding

This method is used for objects such as milk bottle crates, television set cases, safety helmets and construction kits. Poly(ethene) and poly(styrene) are moulded in this way.

Granules of thermoplastic material are fed into a heated cylinder.

4 The plastic cools. The mould is opened, and the finished article is removed

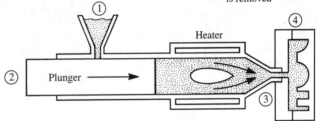

The plunger moves forward.

3 The plunger forces the softened plastic into the unheated mould.

FIGURE 4.15B
Extrusion Moulding

This method is used for making pipes, sheets, fibres and films, according to the shape of the die. It can be used to coat wire, e.g. as insulation for electric cable. PVC pipes and poly(ethene benzene–1,4–dicarboxylate) (PET) fibres and films are made in this way.

1 Granules of thermoplastic are fed into a heated cylinder.

4 The shape of the extruded material depends on the shape of the die.

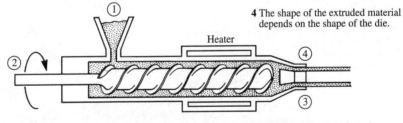

2 A revolving screw moves plastic forward continuously

3 The thermoplastic is forced through a die (a nozzle).

FIGURE 4.15C
Blow Moulding

This method is used for bottles and other containers.

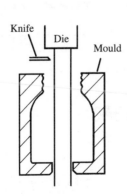
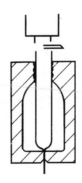
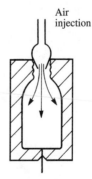

1 Molten polymer is extruded as a hot tube into a suitable mould.

2 The two halves of the mould are closed. The bottom end of the tube is pinched and sealed. The top end is cut by a knife.

3 Compressed air is blown in. The sealed tube is blow up to fill the mould.

4 The mould is opened to release the article.

FIGURE 4.15D
Extrusion Stretch
Blow Moulding

*Methods of moulding
thermosoftening plastics
include ...
... injection moulding ...
... extrusion moulding ...
... blow moulding ...
... extrusion stretch
moulding ...
... vacuum forming ...
... pressure forming ...
... calendering.*

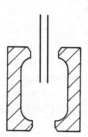

1 Heat-softened
PVC is extruded
as a tube.

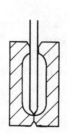

2 A mould closes round a section
of the tube. This mould has a
simple shape called the pre-form.

3 Air is injected to mould the
parison (the section of tube)
into the shape of the pre-form.

4 The pre-form is
transferred into the
strech blow mould.
It is stretched lengthwise.

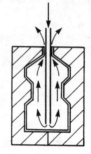

5 Air is injected to blow it
into shape, stretching it
radially.

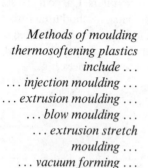

6 The bottle is cooled rapidly
to 'freeze' the molecules in the
streched orientation. It has a
surface area of 4–5 times that
of the pre-form.

FIGURE 4.15E
Vacuum Forming

1. A thin sheet of plastic is softened
   by heating.

2. A mould moves up to make contact with
   the sheet.

3. A vacuum pump is applied to reduce the
   pressure in the space between the sheet
   and the mould. The plastic sheet is
   sucked into the shape of the mould.

4. Later, the mould is moved downwards
   to release the article.

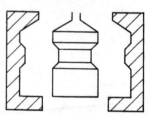

FIGURE 4.15F
Calendering, often used
for PVC

Hopper holds the plastic material.

Heavy heated rollers soften the
plastic and form it into a sheet.

Further rollers polish the sheet
and may emboss or print it.

Finally the sheet is reeled
off through cooling rollers.

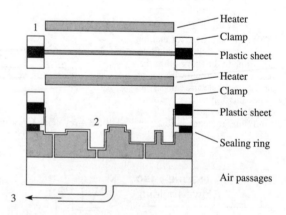

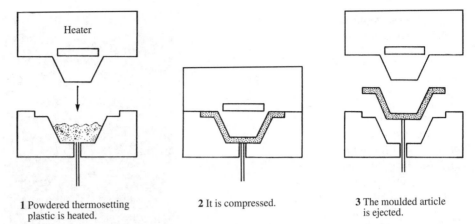

FIGURE 4.15G
Compression Moulding

1 Powdered thermosetting plastic is heated.

2 It is compressed.

3 The moulded article is ejected.

FIGURE 4.15H
Laminating, used for thermosetting resins

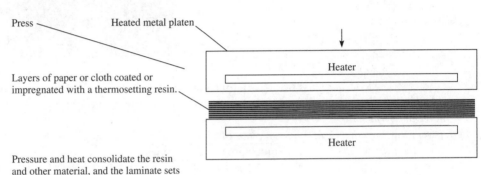

Press — Heated metal platen

Layers of paper or cloth coated or impregnated with a thermosetting resin.

Pressure and heat consolidate the resin and other material, and the laminate sets on cooling.

Methods of moulding thermosetting plastics include . . .

. . . compression moulding . . .

. . . laminating.

Sometimes gases are mixed with plastics during shaping to make plastic foams. These low-density foams, e.g. polystyrene foam, are used in car seats, for thermal insulation of buildings, for insulation against sound and for packaging.

CHECKPOINT 4.15

1. (*a*) State the difference between a batch process and a continuous process.

(*b*) Give two advantages of each type of process as used for carrying out polymerisation.

(*c*) Explain what is meant by mass polymerisation.

(*d*) Why is there a possibility of autoacceleration in mass polymerisation?

(*e*) Why is autoacceleration less likely to happen in solution polymerisation?

2. Mention three types of substance which are added to plastics and say what purpose each achieves.

3. Mention three ways in which the properties of plastics can be modified (other than those mentioned in Question 2). Explain how each of the treatments produces the desired effect.

4. (*a*) Briefly describe the method used to mould (i) a plastic tube, (ii) a plastic bottle.

(*b*) Why can these methods not be used for thermosetting plastics?

4.16 FIBRES

The cables which support a suspension bridge [see Figure 4.16A] are composed of thousands of steel wires intertwined. The cable is stronger than a steel rod of the same diameter; that is, the ratio (tensile strength/diameter) is greater for steel wire than for steel rod. When other materials also are made into fibres, their (tensile strength/diameter) ratio is greater than in bulk form. Glass is a striking example of the difference:

FIGURE 4.16A
A Suspension Bridge

The value of (tensile strength/diameter) is much greater for a fibre than for a rod of the same material.

glass fibre has a much higher (tensile strength/diameter) ratio than glass rod. Some properties of glass fibres were described in § 3.11.2.

A fibre is a piece of material of which the length is very much greater than the width.

A fibre is a piece of material which has a small cross-sectional area and has a length which is much greater than its width.

The materials from which clothing and furnishing fabrics are made are composed of fibres. When a fibre is very long, it is described as a **continuous filament**. If a fibre is less than 10 cm long, it is described as 'staple'. The process of making **yarn** from **staple** is called **spinning**. There are two units for expressing the thickness of a fibre:

> **denier:** the mass in grams of 9000 metres of yarn and

> **tex:** the mass in grams of 1000 metres of yarn.

A fibre may be very long – a continuous filament – or short – staple. Yarn is spun from staple. There are natural fibres, artificial fibres based on cellulose and true synthetic fibres.

In clothing manufacture, the **natural fibres**, cotton and wool, still form 50% of the market in fibres. There are also the **artificial fibres** or **regenerated fibres** which are made from the natural substance cellulose, e.g. viscose and rayon. True **synthetic fibres** originated in 1935 when the British chemist Wallace Carothers made fibres of a **polyamide** and named it **nylon**. The second major type of synthetic fibre to be developed was the **polyesters**, e.g. Terylene. The third type of synthetic fibre, the **acrylics**, were developed by Germany in the 1940s.

4.16.1 MAKING A FIBRE

To form a fibre a polymer molecule must be 10–100 μm long with a relative molecular mass of $1 \times 10^4 - 1 \times 10^6$. ... with strong intermolecular forces.

In order to form a fibre, a polymer molecule must be about 10–100 μm long with a relative molecular mass of 1×10^4–1×10^6. In the case of synthetic fibres, the relative molecular mass is limited by the fact that the molten polymer or a solution of the polymer must have a viscosity that allows it to be spun. Natural fibres have some degree of crystallinity along the fibre axis. Regenerated fibres and synthetic fibres used in the manufacture of textiles need to have a similar structure. These fibres must be drawn or heat-treated to give them a degree of crystallinity. If a fibre is too crystalline, however, it will be too rigid for processing and it will not have disordered regions which can bond to dyes. The strongest fibres have 50–60% crystallinity.

4.16.2 MAKING A FABRIC

To create a material which consists of fibres several stages are involved.

1. *Polymerisation.* The raw material is converted into polymer by addition polymerisation or condensation polymerisation.

2. *Spinning.* The polymer is spun to produce a **filament**, by **melt spinning, wet spinning** or **dry spinning** [see Figure 4.16B].

FIGURE 4.16B
(a) Melt Spinning,
(b) Wet Spinning,
(c) Dry Spinning

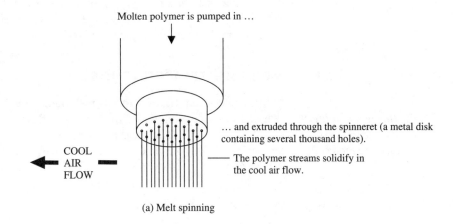

Molten polymer is pumped in …

… and extruded through the spinneret (a metal disk containing several thousand holes).

COOL AIR FLOW

The polymer streams solidify in the cool air flow.

(a) Melt spinning

*Fibres are used to make fabrics. The stages involved are …
… polymerisation of the monomer …
… spinning the monomer into a filament …
… by melt spinning, wet spinning or dry spinning, …
… texturising to introduce air between filaments, …
… weaving or knitting.*

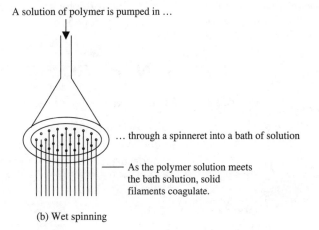

A solution of polymer is pumped in …

… through a spinneret into a bath of solution

As the polymer solution meets the bath solution, solid filaments coagulate.

(b) Wet spinning

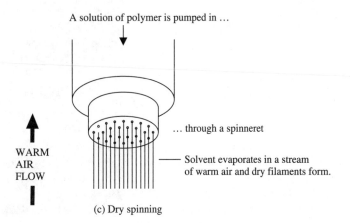

A solution of polymer is pumped in …

… through a spinneret

WARM AIR FLOW

Solvent evaporates in a stream of warm air and dry filaments form.

(c) Dry spinning

3. *Texturising.* The filament is treated to produce a usable **fibre** or **yarn**. After spinning, the filament is **texturised**. Individual filaments are reshaped so that they lie parallel and closely packed but have air space in between them. This texturising makes the eventual yarns bulkier, with a softer feel, improving thermal insulation, moisture absorption and stretch properties. Some filaments are texturised during melt spinning. They may be crimped to create waves along the filament, or loops may be created along the filaments or jets of air under pressure may be used to force bundles of fibres apart and produce the desired fibre. Yarns made from cotton or wool are made by cutting the continuous filaments into shorter lengths. Synthetic yarns can be fabricated in the same way so it is possible to mix natural and synthetic fibres.

4. *Weaving and knitting.* The yarns must be woven or knitted to produce fabrics. Weaving produces fabrics which have tight structures with relatively little stretch. Knitting produces fabrics which need to be close-fitting, e.g. socks, jumpers and tights.

4.16.3   MELTING TEMPERATURE

The melting temperature of a polymer determines the uses which are made of it.

Fabrics made from fibres with low melting temperatures must be treated with care: they can be damaged by hot irons and lighted cigarettes and may lose their shape in very hot water. Poly(ethene) is not used in fibres because of its low melting temperature. Poly(propene) is used in e.g. carpets. It is sold in granule form, heated to 280 °C to give a free-flowing liquid, pigment is added, and then it is melt-spun.

4.16.4   DYEING

New dyes have been made for synthetic fabrics.

Fibres are colourless and their ability to take up dyes is important. Most synthetic fibres are hydrophobic and water-soluble dyes will not bond to them. Many new dyes have been developed for synthetic fabrics.

4.17   SYNTHETIC FIBRES

4.17.1   POLYAMIDES; NYLONS

Nylon stockings were first made in 1939, the year that saw the beginning of the Second World War. Most of the nylon produced at the time was devoted to wartime uses, for instance as a substitute for silk in the manufacture of parachutes, so nylon stockings did not become available until after the end of the war.

FIGURE 4.17A
A Nylon Parachute

About 70% of synthetic fibres are **polyamides** or **polyesters. Nylon** is a general term for polyamide. Nylons contain a relatively large number of crystalline regions arranged in a random manner. If these regions are aligned, there is an increase in the tensile strength [§ 1.4]; see Figure 4.17B.

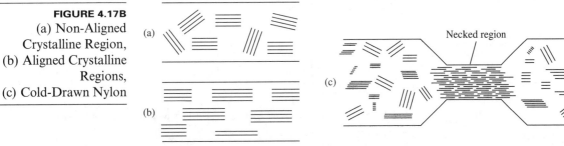

This alignment can be brought about by **cold drawing** – stretching at room temperature. In addition to the tensile strength, it increases the lustre and the resistance to the absorption of moisture. The close packing of nylon chains is one of the factors that prevents penetration by water molecules. Also nylon contains a smaller proportion of —CONH— groups than wool. It also increases the difficulty of dyeing the material. Nylon can be melt-spun, and the fibres are immediately stretched to several times their length [see Figure 4.17C].

Nylon fibres are melt spun.

Nylon is strong, tough and resistant to abrasion and chemical attack. The hydrogen bonds between molecules make nylon stronger than poly(alkenes). The water-repellant quality of nylon makes it useful in swimwear and waterproof jackets, but it also makes nylon clothes unable to absorb perspiration and rather uncomfortable to wear. Nylon fabrics tend to become charged with static electricity. Cotton does not do this because it absorbs water, which conducts electricity and does not allow static to build up. ICI has made an improved nylon named **Tactel** by texturising as described in § 4.16.2 and a delustrant is added to take away some of the shine. The loops give fabrics woven from these fibres a softer texture, more similar to cotton. Further development has produced fabrics made from very fine fibres which allow water vapour to escape between the fibres but do not allow liquid water in. Tactel fibre is used in over half of the skiwear produced in Europe, as well as other sportswear, e.g. jogging suits, tennis outfits, etc.

Nylon fabrics repel water ...
... making them useful in waterproof clothing ...
... but unable to absorb perspiration.
Improvements in nylon have been made by creating loops in the filaments to incorporate air, e.g. Tactel, which allows water vapour to escape but does not allow liquid water to pass.
The strength and durability of nylon find it uses in e.g. conveyor belts and computers.

Nylon has the strength and durability required for use in computer printer ribbons and as conveyor belts.

The UK's worst chemical accident was associated with the manufacture of nylon ...
... and illustrated the danger of a crack in a metal component.

The UK Chemical Industry's Blackest Day

Nylon-6 is made from caprolactam, $HN—(CH_2)_5—C{=}O$ with the group $\overset{O}{\overline{\quad|\quad}}$ The manufacture of caprolactam was responsible for the biggest disaster in the history of the UK chemical industry. It happened in 1974 at the Flixborough works of Nypro (UK) in Humberside. Part of the process involves the oxidation of cyclohexane to cyclohexanol and cyclohexanone. One of the reactors in the plant had been removed for repair and replaced temporarily by a pipe of large diameter. A crack formed in the pipe and cyclohexane leaked out to form a cloud of highly flammable vapour. This ignited. The resulting explosion killed 28 people and injured 100 others as well as damaging 2000 houses and shops. Had the explosion not occurred on a Saturday when few people were working in the plant, the loss of life would have been much greater.

At the time of the accident there was no plant engineer because the post was vacant.

FIGURE 4.17C
Hydrogen Bonding in Nylon

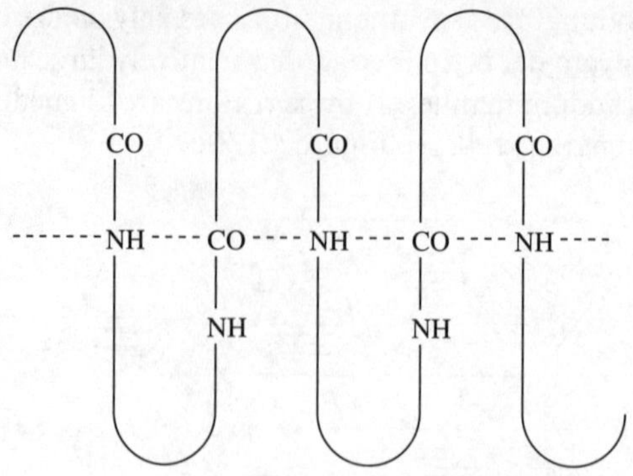

(a) In undrawn nylon, the chains are folded and held in position by hydrogen bonds.

Polyamides or nylons contain crystalline regions arranged in a random manner.
When the fibre is drawn, the crystalline regions are aligned . . .
. . . with an increase in tensile strength.
Nylon chains pack closely together through hydrogen bonding.

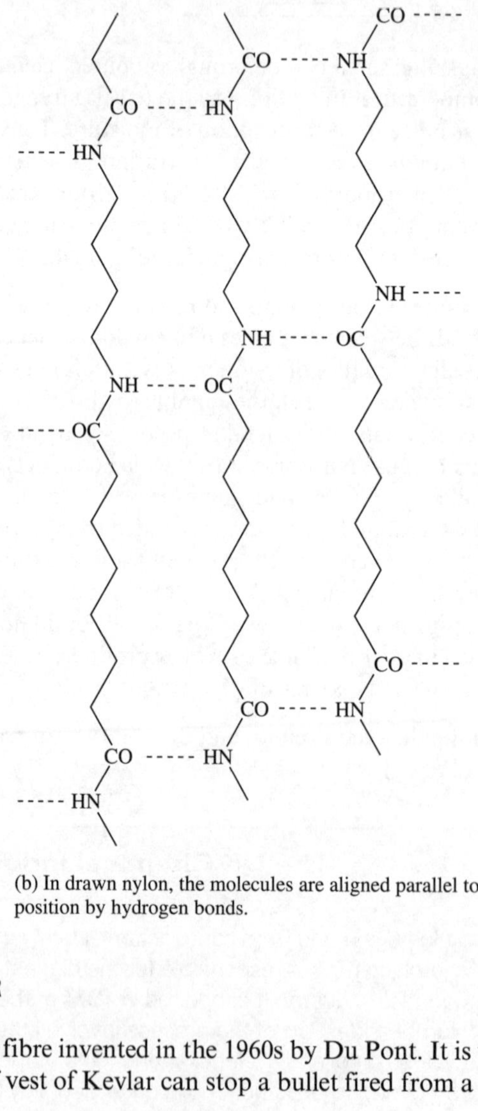

(b) In drawn nylon, the molecules are aligned parallel to one another and held in position by hydrogen bonds.

4.17.2 KEVLAR

Kevlar, is a fibre invented in the 1960s by Du Pont. It is used to make body armour: a bulletproof vest of Kevlar can stop a bullet fired from a handgun at 3 metres.

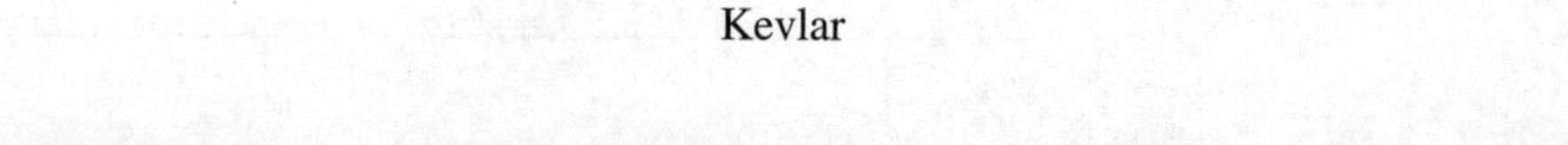

Kevlar

The firm wanted to make a fibre which was as stiff as glass and as heat-resistant as asbestos. The substance they developed, Kevlar, has straight chains which line up parallel to one another and are connected by hydrogen bonds to form sheets of molecules [see Figure 4.17D]. The sheets stack together round the fibre axis to give a highly ordered structure. Kevlar is insoluble in all common solvents except concentrated sulphuric acid, which disrupts the hydrogen bonds between molecules. The fibres are made by wet spinning: extruding a solution of Kevlar in concentrated sulphuric acid into water. Du Pont had to invest in an acid-resistant plant for the process. Kevlar is an **aramid**, an aromatic poly(amide).

Kevlar is used for reinforcing tyres, where it replaces steel wires. It is used to make ropes which are twenty times as strong as steel ropes of the same weight and which last longer. It is used for reinforcing aircraft wings and, as already mentioned, for bullet-proof vests [see also § 5.1].

FIGURE 4.17D
Kevlar. Hydrogen bonds hold the molecules together to form a sheet

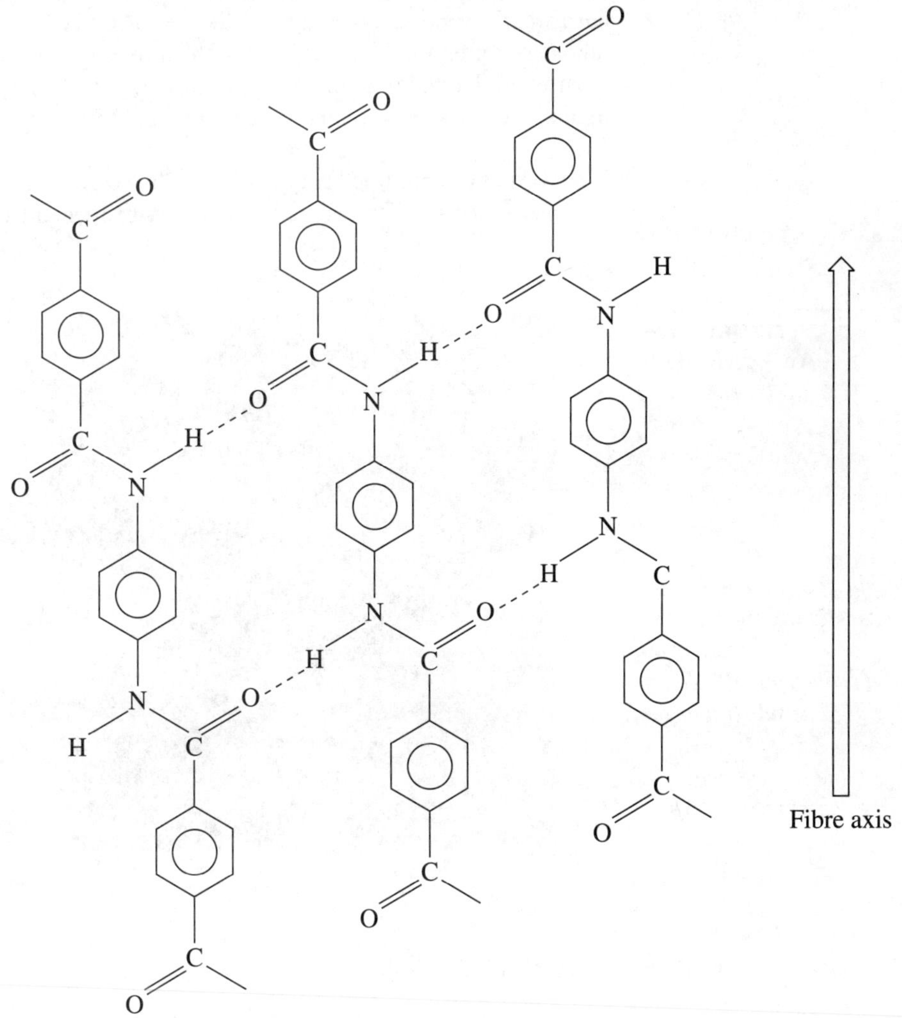

Kevlar is an aromatic polyamide ...
... with strong hydrogen bonding between polymer chains ...
Kevlar fibres are used for reinforcing rubbers and metals.

Fibre axis

4.17.3 POLYESTERS

The chief polyester fabrics used in making garments are Terylene and Dacron, which are trade names for the same polymer.

Many linear polyesters do not make suitable fibres because they have low melting temperatures and high solubilities. In Terylene and Dacron [see § 4.12.1] the benzene rings and the cross-links stiffen the chain and give the polymer a high melting temperature and good fibre-forming qualities. With its high melting temperature, 265 °C, and its high glass transition temperature, Terylene can be melt-spun.

The fabrics are crease-resistant and easy to care for.

Polyesters are strong and remain strong up to 150–200 °C. With crease-resistance and low absorption of moisture, garments made of Terylene resist wrinkling in wear, are quick-drying and need little ironing. Terylene and Dacron are excellent for making trousers and skirts, sheets, boat sails and fillings for duvets and pillows. They can be used alone or blended with cotton to make them better able to absorb moisture.

4.17.4 ACRYLICS AND MODACRYLICS

Poly(propenenitrile), $(CH_2CHCN)_n$, is the polymer of propenenitrile, $CH_2{=}CHCN$, formerly named acrylonitrile. Poly(propenenitrile) can be spun as fibres. DuPont's Orlon was the first fabric to be made from poly(propenenitrile) on a commercial scale. It was followed by Acrilan and Courtelle. These fabrics are called **acrylics**; they are hard-wearing and soft with a wool-like feel. Acrylic fibres are dry-spun and hot-drawn (above room temperature) to stretch them to 3–8 times their original length. Acrylics are used in the manufacture of sweaters, coats, suits, carpets and blankets. They are used in blends with wool and other synthetic fibres.

Modacrylics are modified acrylic fibres. They contain 35% to 85% acrylic fibres. Dynel is a modacrylic; it is made from 40% propenenitrile and 60% chloroethene.

FIGURE 4.17E
An Acrylic Bath

Acrylic fabrics, e.g. Orlon, Acrilan, Courtelle, are made from poly(propenenitrile). They are used as substitutes for wool . . .
. . . in blends with wool . . .
. . . and in modacrylics.

4.17.5 POLYETHERS

A team at ICI in the 1960s tackled the job of making new polymers which could be used at high temperatures; that is, polymers with high melting temperatures and resistance to oxidation. After a good deal of research, they made poly(etheretherketone), PEEK, which has the formula:

The starting materials are

KO—⬡—OK   and   F—⬡—C(=O)—⬡—F

They react with the elimination of potassium fluoride.

Polyethers can be used at high temperatures . . .

PEEK has a high melting temperature, 334 °C and therefore retains good tensile strength and good electrical insulation at temperatures which are high for a polymer. It resists mechanical wear and chemical attack and has low flammability. PEEK is used to insulate cables in environments where toughness and lack of flammability are important.

. . . e.g. poly(etheretherketone), PEEK . . .
. . . which has a high melting temperature . . .
. . . and can be spun to form the fibre Zyex.

PEEK can be spun to produce a fibre called **Zyex**, which is strong and stable at temperatures higher than those tolerated by any other thermoplastic polymer. It is used in industrial applications and in tennis racquets [see § 6.8]. It is better than natural ox-gut for stringing racquets because it is more resistant to abrasion and water and retains its tension over a longer period.

4.17.6   A SUMMARY OF SYNTHETIC FIBRES

The properties and uses of some synthetic fibres are tabulated.

Fibre	Properties	Uses
Nylon	High tensile strength Resistant to abrasion Weakened by acids and by sunlight Low affinity for water Melting temp ≈ 260 °C Can be dyed	Tights, stockings, ropes, parachutes, safety belts, Drive belts, conveyor belts, reinforced rubber tyres, tooth brushes Easy-drying garments Permanent pleats
Terylene, Dacron (Has properties similar to nylon, but is more flammable.)	High tensile strength Resistant to abrasion Melting temp ≈ 230 °C Stable in sunlight Difficult to dye	Ropes, safety belts and harnesses Conveyor and drive belts Permanent pleats Net curtains
Poly(propene) isotactic	High tensile strength Melting temp ≈ 160 °C: easily melted by ironing Water-repellent	Rope, sacks, reinforced concrete, carpets, curtains
Orlon, Acrilan	High tensile strength Less resistant to abrasion than the other fibres Resistant to sunlight and acids, attacked by alkalis Melting temp ≈ 230 °C Can be dyed readily	Sweaters, coats, suits, pile fabrics, e.g. carpets, blends with wool

TABLE 4.17A
Synthetic Fibres

CHECKPOINT 4.17

1. Briefly describe three methods by which a polymer can be spun to produce a filament.

2. What happens when a filament is texturised?

3. Name two methods used to make fabrics from yarns.

4. During the manufacture of nylon fabric, nylon fibres are stretched at room temperature.

(*a*) What name is given to this process?

(*b*) What effect does it have on the arrangement of molecules in the fibre?

(*c*) What advantages does this treatment confer on nylon?

(*d*) Are there any drawbacks?

(*e*) State a disadvantage that nylon has for clothing manufacture compared with cotton.

(*f*) How can the properties of nylon be improved to counter this defect?

(*g*) State an advantage that nylon has over cotton.

(*h*) Name the type of monomer which is polymerised to make nylon.

(*i*) Explain how molecules of nylon bond together strongly.

5. (*a*) Explain why the polymer Kevlar, which has the repeating unit $(COC_6H_4CONHC_6H_4NH)_n$ forms very strong intermolecular bonds.

(*b*) How do the uses to which Kevlar is put depend on the strength of these bonds?

6. Name (*a*) a polyester, (*b*) a poly(propenenitrile) which is used in clothing manufacture. Describe the difference between the two fabrics.

4.18 REGENERATED FIBRES

Regenerated fibres are made from natural materials. A solution of the natural material is extruded through a spinneret into a medium which makes the material solidify in fibre form. **Cellulose** is the most important starting material. It is a polymer of β-glucose.

Regenerated fibres are made from natural materials, e.g. cellulose.

The cellulose in wood contains about 1000 glucose units per molecule, that in cotton 2000 to 10 000 units. The regenerated cellulose fibre, **rayon**, contains 270 glucose units per molecule.

4.18.1 RAYON

Rayon is made from wood pulp, which is soaked in sodium hydroxide solution for several days and then treated with carbon disulphide to form a syrupy solution called **viscose**. Viscose can be wet-spun through a spinneret into sulphuric acid to coagulate the fibres. The fibres are drawn together and stretched and twisted to form a yarn. Variations in the method give rayons of different properties which are put to different uses, e.g. garment fabrics and tyre reinforcement. When viscose is extruded through a slit, sheets of rayon are formed. These are polished to give **cellophane**.

Rayon is made from wood pulp.

Compared with natural cotton, rayon has low wet strength and low water absorbancy and a limp feel. Texturising [see § 4.16.2] has made it possible to make rougher fibres which have a better feel and are crease-resistant.

4.18.2 CELLULOSE ETHANOATE (ACETATE)

Cellulose ethanoate is made from wood pulp and cotton fibres . . .
. . . for use in fabrics, e.g. Dicel and Tricel.

A glucose molecule contains five hydroxyl groups (see formula above). Two are used to form glycosidic links in the polymer cellulose. The remaining three hydroxyl groups can be esterified by ethanoylation (—OH converted into —OCOCH$_3$). High-purity wood pulp and cotton fibres are sources of cellulose. Either material can be soaked in ethanoic acid and allowed to react with a mixture of ethanoic acid and ethanoic anhydride in the presence of sulphuric acid. The solution is wet-spun and the fibres made into fabric. If two hydroxyl groups are ethanoylated, the fabric is called Dicel; if three react the fabric Tricel is formed. Replacing hydroxyl groups by ester groups decreases water absorbancy and improves the feel of the fabric. Tricel is more popular than Dicel because it has a higher melting temperature and lower water absorbancy. Both fibres are used for shirts, blouses and knitwear.

CHECKPOINT 4.18

1. (*a*) What is the difference between a synthetic fibre and a regenerated fibre?

(*b*) Why does the garment industry use more synthetic fibres than regenerated fibres?

2. (*a*) Briefly describe how rayon is made.

(*b*) How does rayon compare with cotton?

3. (*a*) Explain how Dicel and Tricel are made from cellulose.

(*b*) Give examples of their use.

4.19 ELASTOMERS (RUBBERS)

Elastomers are materials that can extend to many times their length. They extend by rotation of covalent bonds in the polymer chain . . .
. . . and contract in order to increase their entropy.

Most materials when stressed to an extension of 1% return to their original dimensions when the stress is removed. When this extension is exceeded, the deformation becomes either non-reversible, as in plastic behaviour, or causes fracture, as in glass and metals. **Elastomers** behave differently: they can extend to up to ten times their length. Elastomers are lightly cross-linked networks, and the structure is able to rearrange by rotation of the covalent bonds in the main polymer chain. When a polymer chain is elongated, the entropy of the system decreases [for entropy, see *ALC*, § 10.9]. The force which retracts the chain is the tendency for the chain to return to the undeformed state of maximum entropy.

Values of the tensile modulus are:

mild steel: 2.2×10^5 Pa, glass: 6×10^4 Pa, polystyrene: 3.2×10^3 Pa
natural rubber: 2 Pa

Rubber particles are incorporated in plastics to stop cracks developing and improve their resistance to impact [see § 1.10].

FIGURE 4.19A
(a) Unstretched Rubber,
(b) Stretched Rubber: the chains have disentangled

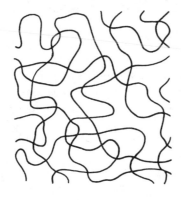
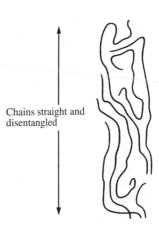

Chains straight and disentangled

Elastomers become glassy below the glass transition temperature . . .
. . . and crystallise when cold-drawn.

4.19.1   ELASTANE, A POLY(URETHANE)

An isocyanate reacts with an alcohol to form a urethane:

$$R'—N{=}C{=}O + HO—R \longrightarrow R'—NH—CO—O—R$$

Isocyanate   Alcohol   Urethane

Poly(urethanes) are formed by reaction between a diiosocyanate and a diol or triol. They are rubbery substances . . .
. . . with resistance to abrasion and solvents . . .
. . . of high tensile strength . . .
. . . and able to withstand high temperatures.

Diisocyanates react with long-chain diols and triols to form **poly(urethanes)** of relative molecular mass 1000–2000. You will probably have carried out this reaction in a lab period. When the liquid diol and the liquid diisocyanate are mixed and stirred, the reaction takes place spontaneously. In a few seconds a mushroom-shaped solid rises out of the beaker. It is hot to the touch because the reaction is strongly exothermic. The polymers are rubbery substances with outstanding properties. They possess high resistance to abrasion and solvents and high tensile strength and can be used at high temperatures.

When the diol contains carboxyl groups as well as hydroxyl groups, the product of condensation is an **elastic fibre**. Gas is evolved in the reaction and serves to expand the mass to yield a **foam**. The product is used as a replacement for foam rubber in upholstery, mattresses, insulation and vibration damping.

FIGURE 4.19B
Photomicrograph of Poly(urethane) Foam

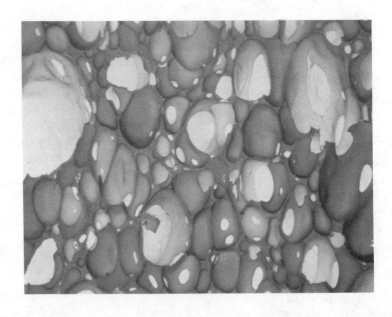

Poly(urethane) foam is used as a substitute for foam rubber.

Elastane is a fibre derived from poly(urethane) which gives the best stretch and recovery performance of any fibre. Elastane is used in swimwear, foundation garments, stockings, tights, etc, under e.g. the brand names Lycra and Spandex.

Elastane, a fibre derived from poly(urethane), has the best stretch and recovery performance of any fibre.

$$\left(\!O—R'—O—\overset{\displaystyle O}{\overset{\|}{C}}—\overset{\displaystyle H}{\overset{|}{N}}—R''—\overset{\displaystyle H}{\overset{|}{N}}—\overset{\displaystyle O}{\overset{\|}{C}}\!\right)_{\!n}$$

4.19.2 NATURAL RUBBER

Natural rubber is a polymer of the monomer 2-methylbuta-1,3-diene (isoprene),

$$
\begin{array}{ccc}
CH_2 & & CH_2 \\
\diagdown\!\!\diagdown & & \diagup\!\!\diagup \\
& C\!-\!C & \\
\diagup & & \diagdown \\
CH_3 & & H
\end{array}
$$

Poly(2-methylbuta-1,3-diene) can exist in two isomeric forms.

Natural rubber is the *cis*-form.

Natural rubber is cis-poly(2-methylbuta-1,3-diene).

The *trans*-form is found in gutta percha, a hard, greyish material which does not change shape and does not resemble rubber.

Gutta percha, which is not elastic, is trans-poly (2-methylbuta-1,3-diene).

The *cis*-structure for poly(isoprene) has the —CH$_3$ groups all on the same side of the chain. The chain can bend and coil in a manner which puts all the CH$_3$ groups on the outside of the bend. This is the structure of natural rubber. In the *trans*-structure, the CH$_3$ groups alternate between opposite sides of the chain, with the result that the chain cannot bend easily because the CH$_3$ groups get in the way of each other. This is the structure of gutta percha, which is inflexible by comparison with rubber.

Rubber is an elastomer ...
... with convoluted molecules ...
... which are straightened out by a force ...
... and contract when the force is removed ...
... but creep can occur.

Rubber is an elastomer which will stretch reversibly to six times its original length. The elasticity of rubber is due to its polymeric structure, the size and shape of the polymer chains and its low glass transition temperature. A rubber consists of a tangle of large numbers of convoluted molecules which can be straightened out if sufficient force is applied. If natural rubber is held in the extended state for some time, it does not contract to its original length; it remains permanently longer. The phenomenon is called **creep** [see § 1.9]. It occurs because polymer chains become separated from one another and slip across one another.

When the maximum amount of creep has taken place, further extension causes fracture of molecular chains, and the rubber breaks. It would obviously be an advantage to eliminate creep, and this can be done by forming cross-links between molecules. In the process of **vulcanisation**, 1–3% by mass of sulphur is added to rubber and the mixture is heated. Short chains of sulphur atoms form between polymer chains.

Vulcanisation creates cross-links and reduces creep ...

Vulcanised rubber is still elastic because when it is stretched the polymer chains can straighten out. The chains cannot creep by slipping across one another since they are joined by covalent bonds. When the stretching force is released, the rubber returns to its original shape. If the extent of vulcanisation is too great, however, the degree of cross-linking reduces the elasticity. Ebonite, a hard, infusible solid with no elastic properties, is made by extensive vulcanisation.

... but allows rubber to stretch.

Vulcanised rubber

ADDITIVES

Additives strengthen and preserve rubber. Tyres utilise most of the rubber produced.

Carbon is added to rubber. It increases the tensile strength, tear strength and resistance to abrasion. Carbon prolongs the life of rubber by absorbing ultraviolet radiation which would otherwise bring about disintegration. Some organic bases are added to remove the free radicals which catalyse the oxidation of rubber by oxygen and also to react with ozone and prevent its attack on double bonds in rubber.

4.19.3 SYNTHETIC RUBBERS

Poly(buta-1,3-diene) has a high resistance to abrasion and retains its elasticity at low temperatures. It can be vulcanised. It is made by polymerising buta-1,3-diene, $CH_2 = CH - CH = CH_2$. Poly(butadienes) are used in the manufacture of tyres, in coatings and adhesives and as additives to other polymers.

Poly(buta-1,3-diene) is a synthetic rubber ... ... which can be vulcanised ... ... copolymerised with phenylethene ... ... and with propenenitrile. Neoprene is similar.

Buta-1,3-diene is often copolymerised with other alkenes to form polymers with a range of properties. With 25% phenylethene (styrene, $C_6H_5CH = CH_2$) it gives styrene–butadiene, SBR, which is produced in larger quantities than any other synthetic rubber and is the only elastomer which is less costly than natural rubber. The largest part of the product is used in tyre manufacture. Carbon black is added to increase the strength, and oil is added as a plasticiser. SBR is also used in shoes and floor tiles, for carpet backing and as an adhesive.

The synthetic rubber NBR is made by copolymerising buta-1,3-diene and propenenitrile, $CH_2 = CHCN$. Neoprene rubber is made from 2-chloro-buta-1,3-diene, $CH_2 = CClCH = CH_2$.

4.19.4 FABRICATION OF RUBBERS

CALENDERING OR COATING

One of the earliest uses for rubber was for coating fabrics to make them waterproof. Rubber is applied to fabrics by calendering – rolling a solution of the rubber into the fabric on calender machines [see Figure 4.15F]. Tyre cord is a special case in which cotton, rayon or polyester cords are arranged in parallel rows and bound together by rubber on a calender.

MOULDING

Rubber compounds can be moulded into any shape, which is retained by curing the compound in the mould. For example, a tennis ball is made by moulding a rubber in the shape of halves of the ball. Two halves are cemented together and vulcanised to form the core of the ball, to which a fabric cover is cemented.

EXTRUDING

Transportation is the major use of rubber, and the tyre is its outstanding product. Inner tubes, tyre treads, weather strips, etc. are made by extrusion [see Figure 4.15B]. A tyre is built up as a cylinder on a drum. Layers of cords and wire cables are incorporated in the tyre and the tread is attached. The drum on which the tyre has been built is then collapsed and the cylindrical tyre is placed in a press. A rubber bag is blown up inside the tyre, and the pressure mould is closed. The tyre becomes doughnut-shaped.

FIGURE 4.19C
Tyre Manufacture

Rubbers are shaped by ...
... calendering a fabric
impregnated with a
solution of rubber ...
... moulding ...
... extrusion, e.g. tyres.
Tennis balls may be hollow
and made of rubber or solid
and made of e.g.
polystyrene

═══════════════════ **CHECKPOINT 4.19** ═══════════════════

1. (*a*) Explain how the structure of rubber enables it to stretch.

(*b*) What makes a rubber contract when the extending force ceases?

(*c*) What happens when the temperature of an elastomer falls below its glass transition temperature?

(*d*) What does vulcanisation do to the structure and properties of rubber?

2. (*a*) Neoprene rubber is made from 2-chloro-buta-1,3-diene, $CH_2=CCl—CH=CH_2$. Write the formula for the repeating unit in Neoprene.

(*b*) The synthetic rubber NBR is made from the monomers buta-1,3-diene, $CH_2=CH—CH=CH_2$, and propenenitrile, $CH_2=CH—CN$. Write the formula of the repeating unit in NBR.

3. Explain why poly(ethene) stands up to bad weather conditions better than rubber does.

4. Natural rubber and gutta percha are isomers of poly(2-methylbuta-1,3-diene). Explain why rubber stretches but gutta percha does not.

4.20 SILICONES

Silicones find a large number of important applications. They are heat-resistant, flexible and electrically insulating. They have the formula (where R = alkyl group or aryl group).

Silicones are polymers that
contain the group R
$$\begin{array}{c} R \\ | \\ —O—Si—O— \\ | \\ R \end{array}$$

$$R—\underset{\underset{R}{|}}{\overset{\overset{R}{|}}{Si}}{\Bigg[}O—\underset{\underset{R}{|}}{\overset{\overset{R}{|}}{Si}}—O{\Bigg]}_n—\underset{\underset{R}{|}}{\overset{\overset{R}{|}}{Si}}—R$$

The size of the polymer molecule determines whether the silicone is a liquid or a viscous resin. Some are cross-linked structures resembling natural silicates.

The value of n determines the properties: when n is small, the polymers are liquids of low viscosity, and when n is large the polymers are viscous gums and resins. Cross-links may be formed by oxygen bridges between siloxane chains. These cross-linked structures resemble natural silicates [see § 3.5] and may consist of long chains (as in asbestos), sheets (as in mica) or three-dimensional tetrahedral structures (as in sand and quartz).

A wide variety of silicone polymers have been made. The properties which make them useful are as follows.

1. The large number of hydrocarbon side-chains, which are hydrophobic, make the material water-repellant.

2. The —Si—O—Si—O—Si—O— structure is thermally stable. With increasing length of side chains the stability decreases.

Silicones find uses because they are:
... water-repellant ...
... stable to heat, ozone, UV radiation ...
... and have viscosity and dielectric properties that do not change much with temperature.

3. Liquid silicones show little change in viscosity as the temperature changes.

4. Silicones resist attack by ozone and ultraviolet light. They are dissolved by organic solvents.

5. Silicones remain electrical insulators over a wide range of temperature, unlike rubber and PVC.

Silicone fluids are used:

Silicone fluids are used as lubricants and water-repellants.

- as lubricants for systems operating at high temperatures, e.g. jet turbine engines
- as water repellants in clothing, tents, furnishing fabrics and coatings for carpets
- as emulsions to waterproof brick, concrete and stone.

Silicone rubbers are used:

Silicone rubbers are used as electrical insulators, for high temperature applications and in medicine.

- in aircraft as ducts to carry gases at high temperature
- in electric insulation
- in industry to produce heat-resistant, non-stick surfaces, e.g. coating rollers and lining pipes for the transport of hot chemicals
- in medicine – research is being done into the use of silicone rubbers in heart valves, transfusion tubing and plastic surgery.

CHECKPOINT 4.20

1. (*a*) What are the outstanding characteristics of silicones?
(*b*) What uses are found for them on the basis of these characteristics?

2. (*a*) How are silicones made?
(*b*) What is the resemblance between silicones and silicates?

4.21 SUMMARY

Table 4.21A lists some polymers and their uses.

Polymer	Properties	Applications
Cellulose esters	Tough. Good impact strength Low thermal conductivity Shiny Thermosoftening	Textile and paper finishes Thickening agents Magnetic tapes Packaging
Epoxy resins	Excellent chemical and thermal stability Strong and tough Electrical insulators Good adhesives Thermosetting	Adhesives Laminates Floorings Surface coatings Linings
Phenoxy resins	Easy to mould, with low shrinkage in mould Self-extinguishing Thermosetting Stable to heat	Surface coatings Adhesives and binders Electronic components
Poly(alkenes)	Excellent resistance to heat, chemicals and water Flexible, with poor mechanical strength Thermosoftening	
Poly(ethene) low density		Films and sheets used in packaging and containers and wire cable insulation
Poly(ethene) high density		Pipes, linings, moulds, coatings, toys, housewares
Poly(propene)	Very strong Hard High melting temperature	Housewares and toys Medical equipment (which can be sterilised) Electronic components Tubes and pipes Fibres and filaments used to make ropes and fishing nets

The table gives a summary of some polymers and their uses.

Polymer	Properties	Applications
Poly(chloroethene) (PVC)	Chemically stable Easy to process Relatively low-cost Self-extinguishing Thermosoftening	Pipes, tubes, gutters, drainpipes, bottles Raincoats, wellingtons Insulation for electrical wiring Floor tiles
Poly(fluoroalkenes), e.g. poly(tetrafluoroethene) (PTFE)	The low coefficient of friction allows few substances to stick to its surface Low permeability Low moisture absorption Exceptional chemical stability Thermosoftening	Electrical insulation Mechanical seals and gaskets. Bearings Linings for chemical equipment Frying pan coatings
Poly(methyl propenoate) (poly(methyl methacrylate)) Perspex	Clear, transparent Good impact strength and tensile strength Thermosoftening	Used in place of glass in vehicle lenses and windows Floor tiles Decorative panels
Poly(phenylethene) (polystyrene)	Low-cost Easily processed Excellent resistance to acids and alkalis but softened by hydrocarbons Clear. Brittle Thermosoftening	Insulation. Packing Vehicle instruments and panels Polystyrene foams are used in containers and as fillings in furniture
Polyamides, e.g. nylon	Tough and strong Easily mouldable Low-density Resistant to abrasion Low coefficient of friction Chemically stable Self-extinguishing Can be drawn into fibres of high tensile strength Thermosoftening	Unlubricated bearings Gears Sutures Fishing lines and nets Tyres Packaging Bottles Ropes Clothing

Polymer	Properties	Applications
Polyesters	Good thermal and chemical stability, though can be hydrolysed by alkali Low-cost Thermosoftening	Vehicle-repair Laminates. Skis Fishing rods. Boats Aircraft parts Bottles
Polycarbonates	High refractive index, transparent Good creep resistance Electrical insulators	Used as replacement for metals in e.g. safety helmets Lenses Insulators Electrical components
Terylene	Electrical insulator Can be drawn into fibres	Fibres used to make textiles which are crease-resistant and easy to wash, dry and iron
Phenolic resins e.g. phenol–methanal, Bakelite	Strong Chemically and thermally stable Good impact strength Electrical insulators High melting temperature Machinable Thermosetting	Glues Laminates Electrical components Switches and sockets Vehicle distributor caps Saucepan handles
Silicones	Stable to heat and oxidisers and water Flexible Electrical insulators Thermosetting	Mould-release agents Rubbers Laminates Antifoaming agents Water-resistant uses
Urea–methanal	Similar to Bakelite but colourless and transparent Can be coloured Thermosetting	Electrical fittings

TABLE 4.21A
A Summary of Uses of
Polymers

4.22 STABILITY OF PLASTICS

A lethal plastic

The piece of plastic which is used to keep the cans together in a six-pack holder is illegal in the USA. It is a piece of poly(ethene), and it does not degrade. Picnickers discard the six-pack holders. Small animals, birds and fish poke their heads through one of the holes, become trapped and drag the plastic round with them for the rest of their lives.

Plastic six-pack holders which degrade – break down in the environment – can be made of a copolymer of ethene and carbon monoxide. Ethene is polymerised with a small amount of carbon monoxide. The carbonyl groups in the polymer absorb some of the sunlight which falls upon them. The energy is used to break bonds in the polymer and degrade it. Poly(ethene) containing 1% of carbonyl groups loses its strength after 32 days in sunlight. Without carbonyl groups, it takes 300 days of sunlight.

Plastics which do not degrade can pose a threat to animals.

HEAT   *disadvantage*

Polymers do not, in general, have high melting temperatures, and this restricts their use. High temperatures must be avoided in the methods used for the processing of polymers.

Polymers are susceptible to attack by . . .
. . . heat – because they have low melting temperatures . . .

AIR   *Not PVC*

Poly(dienes), e.g. poly(buta-1,3-diene) contain double bonds and these can react with the oxygen in the air. Such polymers are stabilised by the addition of antioxidants which act by trapping free radicals. Carbon black is added to stabilise plastics against the absorption of ultraviolet radiation.

. . . light – because it breaks some bonds . . .
. . . air – because oxygen adds to double bonds . . .

CHEMICAL ATTACK   *ad*

Polyamides and polyesters are vulnerable to hydrolysis by acids and alkalis. Organic polymers can dissolve in organic solvents. Plastics have more resistance to attack by acids and by water than metals have. Ceramics are less chemically reactive than plastics are.

BIODEGRADABILITY

Natural polymers, e.g. wood and paper, are biodegradable because micro-organisms in water and in the soil use them as food. For synthetic polymers there are no corresponding micro-organisms, and these non-biodegradable materials can remain in the environment for a very long time.

. . . chemicals – because polyamides and polyesters are hydrolysable . . .
. . . micro-organisms – which digest natural polymers but not synthetic polymers . . .
. . . fire – as many polymers are flammable.

FIRE HAZARD   *not PVC*

Chairs and settees stuffed with poly(urethane) can fill a room with lethal smoke within 2 minutes. As a result of some fatal accidents, furniture manufacturers have now started using other filling materials. One solution to the problem of fire is the use of aluminium oxide-3-water, known as aluminium trihydrate, ATH, as a fire-retardant filler. A mixture of ATH and methyl 2-methylpropenoate, $(CH_3)_2CHCO_2CH_3$, polymerises to give a material with the trade name of **Avron**. This plastic is used for table tops, for floors and for bar tops: it is not damaged by burning cigarettes. ATH is also added to **Modar** (modified propenoic resin or modified acrylic resin).

4.23 DISPOSAL OF PLASTICS

Plastic waste constitutes about 7% of household waste. Unlike some other wastes, c.g. kitchen waste and paper, plastics are non-biodegradable. Plastic waste is buried in landfill sites, and there it remains unchanged for decades. Local authorities have to find more and more landfill sites.

4.23.1 INCINERATION OF PLASTICS

A problem with most plastics is non-biodegradability ...

... and the disposal of plastic waste occupies more and more landfill sites.

An alternative to dumping is **incineration**, with the possibility of making use of the heat generated. Plastics are petroleum products, and plastic waste contains about the same amount of energy as the oil from which it came. To burn plastic waste with the release of useful energy is an obvious solution to the problem.

Incineration of plastics releases energy which can be utilised.

The UK incinerates only about 2% of its municipal solid waste with recovery of the heat evolved and makes use of only about 6% of its plastic waste. In some other countries, e.g. Denmark and Japan, incinerators consume over 70% of domestic waste. The plastic part of the waste assists in the incineration of other parts of domestic rubbish. If plastics are removed from domestic waste, together with paper, what is left is organic waste which is too wet to burn. If the waste is burnt with the plastics included, potentially useful energy is generated. Some plastics, however, burn with the formation of toxic gases, e.g. hydrogen chloride, from PVC, and hydrogen cyanide, from poly(propenenitrile), and incinerators must be designed to remove these gases from the exhaust.

4.23.2 RECYCLING PLASTICS

Recycling is a very efficient method of dealing with plastics waste.

A difficulty is that plastics waste is a mixture of plastics of different properties.

The EC is urging member nations to recycle used materials, and the UK Government has set a target of 50% recycling of household waste by the year 2000. Recycling is the most efficient use of resources. The municipal solid waste collected in the UK each year contains about 1.5 million tonnes of plastics. In addition, used plastics from large articles such as fridges, washing machines, agricultural machinery, cars, etc. bring the total up to about 1.8 million tonnes. A major difficulty in recycling is that, since different plastics have widely differing properties, mixed plastic waste is of limited use. While any two glass bottles can be recycled together, the same is not true of a PVC bottle and a poly(ethene) bottle. If these are melted down together, it may be possible to make some articles from the mixture, but mixtures of plastics are much weaker than individual plastics.

The easiest plastics to recycle are off-cuts and substandard products, which consist of a known single plastic. The average car contains about 140 kg of plastics, which end up as waste when the vehicle reaches the end of its life. Refrigerators and washing machines pose a similar problem, Manufacturers are considering printing bar codes on plastic components to identify them and assist in sorting them into different types of plastic waste.

Individual plastics can be recycled with ease. Some plastic waste is separated by hand ...

... and a start has been made on mechanical methods of separation.

Separating the different plastics makes recycling more worthwhile. The sorting of plastics into individual types is an enormous task. Some sorting is done by hand: plastic bottles can be sorted into poly(ethene), poly(chloroethene) (PVC) and poly(ethene benzene-1,4-dicarboxylate) (PET). Mechanical separation is difficult but a process has been devised which detects the chlorine in PVC by means of X-rays, allowing PVC bottles to be separated from poly(ethene) and PET. These are then separated on the basis of their different densities. Recycled PVC is used for drains and sewer pipes, shoe soles, flooring and packaging of non-food items.

In cases where collection and separation are easy, recycling is profitable. Examples are poly(ethene) from agricultural mulch films (films with holes through which the crop can

In some cases plastic items can be collected separately for recycling.

grow and which must rot or be removed at the end of the growing season) and from shrink-wrap packaging from supermarkets. About 10% of the poly(ethene) film we use is recycled and used to make black refuse bags etc. Telephone hand sets of poly(ethene) are collected and recycled. Poly(propene) car bumpers and casings from car batteries are recycled. Nevertheless only about 7% of the poly(propene) we use is recycled. Soft-drink bottles made of PET can be melted down and moulded into new bottles. They can also be melted down and drawn into fibres to be used in cushion filling and upholstery stuffing and carpet manufacture. All these items can be collected separately. Fast food counters use poly(styrene) containers because they offer thermal insulation. There is a UK poly(styrene) recycling organisation which targets fast food outlets.

Some mixed plastics waste can be melted and remoulded.
A combination of incineration and recycling could make use of a higher fraction of plastic waste.

With municipal solid waste, it has been estimated that no more than 50% of the plastic material can be separated, cleaned and reprocessed. Some of the mixed plastics recovered from household waste is recycled by shredding, melting and extrusion in the shape of planks. This recycled plastic can be used to construct items such as agricultural fencing, pigsties, pens and garden seats. There is a limited demand for plastics of this quality.

4.23.3 PYROLYSIS

Pyrolysis of plastics produces useful substances.
A solution to the problem of discarded tyres is described.

When plastics are heated in air, they burn. When they are heated in the absence of air, they are **pyrolysed** (split up by heat). The products can be separated by fractional distillation and then used in the manufacture of other materials, including plastics. The heat generated by burning a small proportion of a load of plastics waste can be used to crack the rest of the load in the furnace. Problems still to be solved include the difficulty of separating different types of plastics, and the difficulty of removing additives. The process is expensive and at present costs about five times as much as dumping the plastics in a landfill site.

4.23.4 BIODEGRADABLE PLASTICS

Chemists have invented some biodegradable plastics. These may be **biopolymers**, which are made by living organisms or **synthetic plastics**. Three types of synthetic plastics are **photodegradable plastics, synthetic biodegradable plastics** and **water-soluble plastics**.

BIOPOLYMERS

Biodegradable plastics have been developed.
Biopolymers are made from natural products and can be broken down by micro-organisms.

In the UK, ICI markets the biopolymer poly(3-hydroxybutanoic acid), PHB, which has the trade name Biopol. It is made by certain bacteria from glucose. Biopol is used for special applications such as surgical stitches which dissolve in time inside the body. There is another potential use for Biopol. When Biopol is discarded, micro-organisms in the soil, in river water and in the body can break it down within 9 months. By incorporating copolymers, the properties of the plastic can be tailored to make it suitable for a range of articles, e.g. shampoo bottles and carrier bags. At present a Biopol container is seven times the price of a poly(ethene) container, but if there were a decision to use biodegradable containers, the price of Biopol would fall.

PHOTODEGRADABLE PLASTICS

Photodegradable plastics break down in sunlight

A Canadian firm has produced a photodegradable polymer, which they incorporate in polystyrene cups. Exposed to sunlight for 60 days, the cups break down into dust particles. To make plastics photodegradable, it is necessary to incorporate in them a substance that will absorb sunlight and as a result become sufficiently reactive to react with plastic molecules. Alternatively, if a carbonyl group can be incorporated into the polymer chain, the carbonyl group will absorb light and use the energy to break chemical bonds in the polymer. Introduction of carbonyl groups is done by polymerising the monomer, e.g. ethene, with carbon monoxide. Polymers containing 1% of carbonyl groups lose their strength after 2 days of sunlight, compared with about 300 days in the absence of carbonyl groups.

SYNTHETIC BIODEGRADABLE PLASTICS

Synthetic biodegradable plastics incorporate natural materials, e.g. starch, in the plastic so that the natural material will be digested by micro-organisms.

An Italian company, Feruzzi, has produced a biodegradable polymer which is suitable for carrier bags. The material consists of poly(ethene) and up to 50% starch. The poly(ethene) chains and starch chains interweave to form a material which is strong enough for shopping bags. When the material is buried, micro-organisms begin to feed on the starch, converting it into carbon dioxide and water, and in time the polymer chains dissolve in water. The material is an example of a polymer alloy [see §4.19]. The cost at present is about twice that of a regular plastic bag. Europe uses 100 000 tonnes of degradable plastic a year, and the consumption is rising.

SOLUBLE PLASTICS

Soluble plastics can be designed to dissolve slowly in hot water, warm water or cold water.

Plastics which dissolve in water can be designed to be soluble in cold water, in warm water or only in hot water. Poly(ethenol), $(\text{—CH}_2\text{CHOH—})_n$, also called poly(vinyl alcohol), PVA, is an example. It is used as a packaging for swimming pool chemicals, descalers, seed strips and other uses to a total of 100 tonnes a year in the UK.

All at sea

A cargo ship had a crew of 50 sailors. During a 6 week voyage the crew threw overboard 350 cardboard boxes, 200 crisp packets, 20 plastic bags, 250 bottles, 2 plastic drums, 2 metal drums, 20 fluorescent tubes, 380 plastic beer can holders and 5200 cans.

The International Convention for the Prevention of Pollution from Ships Annex V came into force in 1989 and outlaws this kind of behaviour. It has been accepted by 41 nations, including the UK, but the signatories to the Annex sail only 60% of the world's ships. Annex V regulates the disposal of rubbish and outlaws the discharge of plastic waste at sea. Until Annex V is signed by all the nations, the problem of plastic waste can best be solved by degradable plastics.

4.23.5 RECYCLING VERSUS BIODEGRADABILITY

The governments of the USA, Sweden and Italy have passed laws making degradability compulsory for plastics in certain types of packaging. On the other hand, Friends of the Earth and the British Plastics Federation now oppose degradable plastics on the grounds that it is better to recycle plastic waste.

Some plastic waste is never collected, e.g. plastics used for agricultural purposes, such as baler twine, bird netting, sapling protectors, mulch films and also plastics thrown overboard from ships. There is much to be gained from using biodegradable plastics for such articles. Other plastic objects could be collected for recycling. Some plastics are easy to recycle, e.g. poly(ethene benzene-1,4-dicarboxylate), PET, from which many bottles and jars are made. There are problems in recycling plastics, however, as described earlier.

There is debate on the merits of biodegradable plastics versus recyclable plastics.

There are two solutions to the problem of plastic waste. One is to make degradable plastics, and the other is to recycle plastics. The two solutions do not live well together. Although some biodegradable plastics, e.g. PHB, can be recycled with other plastics, photodegradable plastics cannot be included. Waste plastics can be turned into items such as sacks, park benches, roofing and drain-pipes. You can imagine the accidents that could occur if such materials were to break up in sunlight.

══════════════════════════ CHECKPOINT 4.23 ══════════════════════════

1. Millions of plastic bags are discarded after one or two hours' use. Many plastic bags are made of poly(ethene).

(*a*) Explain how poly(ethene) is obtained from petroleum.

(*b*) How long did petroleum take to form?

(*c*) Can it be replaced?

(*d*) What is meant by the statement that plastic bags are non-biodegradable? What significance does this statement have for the disposal of plastic waste?

2. Gas is used to convert polystyrene into polystyrene foam.

(*a*) What is the advantage of polystyrene foam for serving food in take-away restaurants?

(*b*) For how long is a polystyrene foam package in use?

(*c*) What happens to the gas it contains?

3. (*a*) Explain the advantage of using a dissolving polymer for (i) surgical stitches, (ii) a laundry bag used to store laundry in a hospital where there is danger of infection.

(*b*) Give three uses for which a dissolving plastic would be unsuitable.

4. Suggest (*a*) applications for a photodegradable plastic, (*b*) items for which it would not be suitable.

5. Discuss the pros and cons of recycling plastics or using biodegradable plastics.

4.24 NOMENCLATURE

While chemists have used the IUPAC systematic names [see *ALC*, §§ 25.4 and 28.2] for organic compounds, manufacturers still use traditional names, and it is not obvious why PVC should be the trade name of poly(chloroethene). The following list of systematic names and traditional names for plastics and related compounds may be useful.

You need to be bilingual! In addition to the systematic names which you have learned, industrialists use traditional names, such as styrene and acrylonitrile. The most common plastics are known by initials, e.g. PVC and PET, which stand for their traditional

Systematic name	Traditional name
Butanoate	Butyrate
Butanoic acid	Butyric acid
Carbamide	Urea
2-Chlorobuta-1,3-diene	Chloroprene
2-Methylbuta-1,3-diene	Isoprene
2-Methylpropenoic acid	Methacrylic acid
Benzene-1,4-dicarboxylic acid	Terephthalic acid
Chloroethene	Vinyl chloride
Ethanoate	Acetate
Ethanoic acid	Acetic acid
Ethene	Ethylene
Ethenol	Vinyl alcohol
Ethenyl ethanoate	Vinyl acetate
Ethenyl group	Vinyl group
Hexanedioic acid	Adipic acid
Hexanoic acid	Caproic acid
Methanal	Formaldehyde
Methyl 2-methylpropenoate	Methyl methacrylate
Methylbuta-1,3-diene	Isoprene
Phenylethene	Styrene
Propene	Propylene
Propenoic acid	Acrylic acid
Propenoic resin	Acrylic resin
Propenenitrile	Acrylonitrile

Table 4.24A
Systematic Names and
Traditional Names

QUESTIONS ON CHAPTER 4

1. PHB stands for poly(hydroxybutanoate).

(*a*) Draw the structural formula of 4-hydroxybutanoic acid.

(*b*) The —CO_2H group of one molecule forms an ester with the —OH group of a second molecule. Draw the structural formula of this ester.

(*c*) Draw the repeating unit in a long chain of poly(hydroxybutanoate).

2. (*a*) Write the structural formulae of (i) hexanedioic acid, (ii) hexanedioyl chloride, (iii) hexane-1,6-diamine.

(*b*) Write the structural formula of the repeating unit in nylon 6,6, which is formed by condensation polymerisation of (ii) and (iii) in part (*a*).

(*c*) The nylon produced in polymerisation contains randomly arranged crystalline regions. What is done to change the random arrangement of these regions in nylon which is to be used as fibres? What beneficial effect does this treatment have on the properties of nylon? What disadvantage does this treatment have on nylon as a fibre for use in the manufacture of clothing?

(*d*) Give two reasons why nylon has poor water absorbancy. How does this affect the use of nylon in the clothing industry?

3. (*a*) What raw materials are used for the manufacture of rayon?

(*b*) Sketch the structure of rayon.

(*c*) Compare rayon and cotton for use in the clothing industry.

4. (*a*) Draw structural formulae showing the three types of structure for poly(propene). Give the name of each structure. State the difference in properties between them.

(*b*) Describe the structural differences between low-density poly(ethene) and high-density poly(ethene).

(*c*) Describe the differences in the manufacturing processes for the two types of poly(ethene).

5. (*a*) Explain how the form of a polymer molecular chain determines the degree of crystallinity which is possible in a polymer. Give examples.

(*b*) Explain how the crystallinity of a polymer affects its properties. Give examples.

6. (*a*) PVC has a glass transition temperature of 87 °C. How do the properties below this temperature differ from those above it?

(*b*) Why are the properties of a cold-drawn polymer different from those of the undrawn polymer?

(*c*) Why are polyester fibres cold-drawn before use?

(*d*) When a piece of poly(ethene) is drawn it starts necking at one point. Further drawing results in no further reduction in cross-sectional area at the necked region but a spread of the necked region along the whole length of the material. Why does the material not simply break at the necked region instead of extending the necking?

7. Poly(ethene) is made by addition polymerisation of ethene.

(*a*) What raw material is used for the production of ethene?

(*b*) Outline the mechanism of polymerisation.

8. (*a*) Explain what is meant by the terms (i) cross-linked polymer, (ii) copolymer.

(*b*) You are given rods of three solids, a polymer of A, a polymer of B and a copolymer of A and B. Outline how you could compare the three specimens as regards (i) density, (ii) tensile strength, (iii) tensile modulus, (iv) impact strength, (v) hardness [see Chapter 1 if necessary].

9. (*a*) Explain what is meant by (i) a thermoplastic, (ii) a thermosetting plastic. Give two examples of each.

(*b*) What is the effect of a plasticiser on the mechanical properties of a polymer?

(*c*) Outline the manufacture of a commercially important polymer.

10. (*a*) What is the vulcanisation of rubber?

(*b*) State three ways in which vulcanisation improves the properties of rubber for use in car tyres.

(*c*) Explain why the vulcanisation of synthetic rubber is unnecessary.

11. Consider what happens when you stretch a strip of poly(ethene), e.g. from a plastic bag. When you stretch it rapidly it breaks, but when you stretch it slowly it elongates until it has reached a length much greater than its original length before finally breaking.

12.

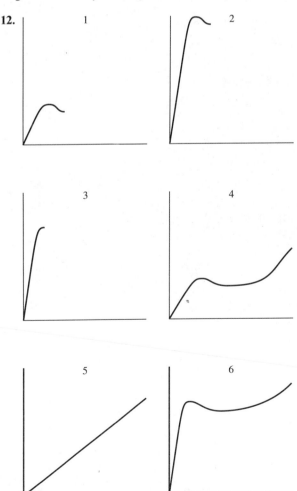

The diagrams 1–6 are stress–strain graphs for different types of polymers. The polymers can be described as

A hard and tough

B hard and brittle

C hard and strong

D soft and weak

E soft and tough

F an elastomer

Match each diagram to the type of polymer. [See Chapter 1 if necessary].

13. Someone gives you a sample of material which he says is suitable for use in the manufacture of 13 A mains electric plugs. He asks you to assess the suitability of the material.

(*a*) After examining a plug, say what properties a plug must have.

(*b*) What tests would you do on the sample material to find out whether it is suitable?

(*c*) List other materials that are suitable for use.

(*d*) What process could be used for moulding the sample material?

14. The formulae of some polymers are given.

A $\displaystyle \cdot\!\!\left(CO(CH_2)_4CONH(CH_2)_6NH\right)\!\!_n$

B $\displaystyle \cdot\!\!\left(COC_6H_4CO_2CH_2CH_2O\right)\!\!_n$

C $\displaystyle \cdot\!\!\left(CH_2\!\!-\!\!CH(CH_3)\right)\!\!_n$

D $\displaystyle \cdot\!\!\left(CH_2\!\!-\!\!C(CH_3)\right)\!\!_n$
$$CH\!\!=\!\!CH_2$$

E $\displaystyle \cdot\!\!\left(CH_2\!\!-\!\!CH\right)\!\!_n$
$$C_6H_5$$

F $\displaystyle \cdot\!\!\left(CH_2\!\!-\!\!CHCl\right)\!\!_n$

G $$R$$
$$\cdot\!\!\left(Si\!\!-\!\!O\right)\!\!_n$$
$$R$$

H $\displaystyle \cdot\!\!\left(CH(R)CONH\right)\!\!_n$

I $\displaystyle \cdot\!\!\left(CH_2CH_2O_2CCH\!\!=\!\!CHCO_2\right)\!\!_n$

J $\displaystyle \cdot\!\!\left(C_6H_4NHCOC_6H_4CONH\right)\!\!_n$

K $\displaystyle \cdot\!\!\left(O\!\!-\!\!C_6H_4\!\!-\!\!O\!\!-\!\!C_6H_4\!\!-\!\!CO\!\!-\!\!C_6H_4\right)\!\!_n$

For each polymer, say (*a*) whether it forms hydrogen bonds, (*b*) whether it can be hydrolysed, (*c*) what type of polymer it is, e.g. poly(alkene), and (*d*) any other information you can offer.

5

COMPOSITE MATERIALS

5.1 BULLET-PROOF VESTS

Bullet-proof vests need to have the ability to absorb and disperse the impact energy of a bullet [see Figure 5.1A]. This ability is achieved in a material which consists of a polymer resin reinforced with fibres of high molar mass poly(ethene). The fibres have a high specific modulus (elastic modulus/density). The structure consists of layers of aligned fibres. The direction of the fibres is rotated through 90° in alternate layers [Figure 5.1B]. When alternate layers are built up continuously a rigid armour suitable for vehicles and riot shields is obtained. For a ballistic vest, greater flexibility is required and this is achieved by sandwiching alternate layers of fibre-reinforced resin between films of low-density poly(ethene).

A bullet-proof vest owes its strength to its composite structure.

FIGURE 5.1A
A Bullet Being Stopped. See how the bullet is flattened as its energy is transferred to the fibres in contact with it.

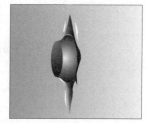

FIGURE 5.1B
The Arrangement of Fibres of Poly(ethene) in a Resin Matrix

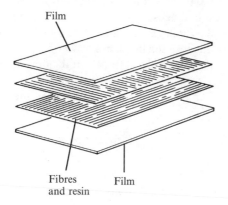
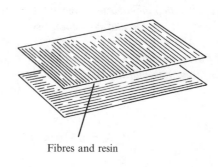

Film

Fibres and resin

Fibres and resin   Film

5.2 TYPES OF COMPOSITE MATERIALS

The materials used to make ballistic armour and ballistic vests are described as composite materials. Neither the resin nor the fibres alone could achieve the required properties, but the combination does the job. A **composite material** is one that is composed of two or more different materials bonded together with one serving as the **matrix** surrounding **particles** or **fibres** of the other. In a composite material, the properties of the components combine to give a material which is more useful for a particular purpose than the individual components.

Another example is reinforced concrete, which has steel rods embedded in the concrete. The composite can carry loads that could not be carried by concrete alone. Concrete itself is a composite: it consists of coarse particles (stone chips or gravel) in a matrix made by mixing cement, sand and water.

In a composite material . . .
. . . one material is the matrix . . .
. . . in which particles or fibres of the second material are embedded.

Many plastics are reinforced with glass fibres or glass particles. Vehicle tyres are made of rubber reinforced with woven cords. Wood consists of tubes of cellulose bonded by a natural plastic called lignin [see Figure 5.10B]. **Cermets**, which are used in the tips of cutting tools, have *cer*amic particles in a *met*al matrix (*cer* + *met* = cermet).

Composites can be:

- **particulate.** The phase of high tensile strength in the form of particles is dispersed in a matrix of the second phase.
- **fibrous.** The strong phase in fibre form is in a matrix of the second phase. The fibres may be continuous or in short lengths. They may be aligned or randomly oriented.
- **laminated.** There are alternating layers of strong phase and weaker phase.

Composites combine the properties of their components.
They can be particulate . . .
. . . or fibrous
. . . or laminated.

The plastic matrices in composites may be thermosetting or thermoplastic. Thermosets are more suited to combination with fibres because they start as a viscous liquid resin which can penetrate the fibre bundles and can be cured (converted by the application of heat) into rigid form. Epoxy resins [4.12.5] and polyester resins [4.12.2] are chiefly used in the manufacture of composites.

5.3 FIBRE-REINFORCED COMPOSITES

The fibres in a composite carry most of the load and provide stiffness.
They must have high tensile strength and a high tensile modulus.
The matrix must adhere to the fibres, protect the fibre surfaces and separate the fibres.

There are three components in a structural composite material: reinforcing fibres, a matrix or glue to stick the fibres together and the interface or interphase between the two. The functions of the fibres in a composite are to carry most of the load applied to the composite and to provide stiffness. The fibrous materials used must have high tensile strength and high tensile modulus. Ceramics are often used as the fibres in composites: as well as high tensile strength and high tensile modulus they have low density. A disadvantage is the brittleness of ceramics and the presence of small flaws which can reduce the tensile strength. When the fibres are incorporated in a ductile matrix, they benefit from the protective properties of the matrix. The composite has properties better than either the matrix material or imperfect fibres.

5.3.1 FIBRES

Fibres make up 10–70% of a composite . . .
. . . and may be short or continuous . . .
. . . may be aligned or randomly oriented.

The fibre content of composites varies from 10 to 70% by volume. The fibres may be discontinuous (short, about 0.25 mm) or continuous fibres with lengths running the full length of the composite. They may be aligned so that they are all in the same direction or they may be randomly oriented. Continuous fibres give the highest tensile strength and tensile modulus but they also give a direction to the properties: the strength along the direction of the fibres could be 800 MPa, while that at right angles to the fibre direction could be only 30 MPa, not much greater than that of the plastic alone. Randomly oriented short fibres avoid this directionality of properties, but they do not give such high tensile strength and tensile modulus.

The development of new and improved fibres has been the main driving force behind the development of new composite materials. By drawing materials such as glass and ceramics [§ 4.16] into fine fibres (say 10 μm diameter), faults within their structure, which might otherwise weaken them, can be eliminated. Heat treatment is another method of enhancing the properties of fibres.

CARBON FIBRES

These consist of bundles of long chains of carbon atoms. They are made by the incomplete combustion of polymer fibres, e.g. poly(propenenitrile). These are stretched during combustion so that highly oriented and therefore strong fibres, consisting of long chains of carbon atoms, are produced. The chains of carbon atoms are joined by cross-links to form a ladder-like structure of great strength and stability. A carbon fibre is a bundle of intertwined ladder-chains. It may have the same stiffness and strength as steel with a much lower density. Although the tensile strength of carbon fibres is not high, because of their stiffness they are included in materials used for the frames of aircraft.

GLASS FIBRES

Glass fibres and ceramic fibres are strong materials. Strength and stiffness are increased . . .

These are heavier than carbon fibres but they are expensive. Fibre glass has been widely used for many years, e.g. in the construction of small boats and surfboards. A recent development is E-glass fibres which are very stiff.

KEVLAR FIBRES

. . . by drawing . . .
. . . and by heat treatment.

Kevlar fibres, $-(C_6H_4CONH-)_n$, are five times as strong as steel, of low density and flexible. These fibres are used where toughness is essential. They are used in the manufacture of tennis racquets for professionals and to strengthen car tyres.

CERAMIC FIBRES

Ceramic fibres, e.g. silicon carbide and aluminium oxide, are used in high-tech tennis racquets. They are claimed to have good impact-absorbing characteristics and therefore to help in avoiding injuries such as tennis elbow.

WHISKERS

Whiskers are exceptionally strong.

Fibres grown from a single crystal are called whiskers and have exceptionally high tensile strength, e.g. alumina whiskers and graphite whiskers.

Property	*Carbon*	*Kevlar*	*E-glass*	*SiC*	*Poly (ethene)*
Fibre diameter/mm	1×10^{-4}	2×10^{-4}	2×10^{-4}	1.4×10^{-4}	5×10^{-6}
Density/Mg m^{-3}	1.80	1.44	2.54	3.00	0.97
Tensile modulus/MN m^{-2}	2.2×10^5	1.2×10^5	7.2×10^5	4.0×10^5	1.2×10^5
Tensile strength/MN m^{-2}	2070	2760	3450	4830	2590
Elongation to fracture/%	1.2	2.4	4.8	—	3.8

TABLE 5.3A
The Properties of some Fibres used in Composites

5.3.2 MATRIX MATERIALS

The functions of the **matrix materials** are to support the fibres, to transfer the load between neighbouring fibres and also to define the shape of the material. Metals, polymers, ceramics and carbon are chosen as matrices depending on the performance required. Polymers have melting temperatures up to 475 K; carbon-fibre-reinforced carbon can be used up to 2275 K in e.g. rocket nozzles.

The commonest matrix in which the fibres are incorporated is an epoxy resin [see §4.12.5]. Thermosetting plastics, such as epoxy resins and polyesters can be moulded once only: when set, they harden and become quite brittle. However, the fibres make up for this brittleness. The fibres are woven into the epoxy glue while it is still viscous and shaped into the required 'preform' which is then hardened. An example is.**carbon-fibre-reinforced epoxy resin:** this is used in tennis racquets and in aircraft. It is stronger and lower in density than conventional materials, e.g. aluminium alloys. Cost rules out the use of carbon fibre composites in place of steel and concrete for large structures, but in articles such as tennis racquets and critical components of aircraft frames carbon fibre composites have the edge over metals.

Polyesters are also reinforced with fibres. The polyester matrix is flexible but weak. The glass fibres are extremely strong, but relatively brittle. The combination of the two materials in **glass-fibre-reinforced-polyesters** produces tough, strong materials. The composites have many applications, including small boats, skis and motor vehicle bodies.

FIGURE 5.3A
Glass-Fibre-Reinforced
Polyester (GRP)

Matrix materials support the fibres and give a shape to the material.

PEEK (poly-ether-ether-ketone) is a recently developed matrix. It is tougher than other binders and more expensive than epoxy resin but less brittle and likely to crack. PEEK is excellent for covering electrical wires, resisting cuts and breaks at sharp corners. The high cost limits the use of PEEK at present. Mouldings of PEEK are used in harsh environments such as nuclear plants, oil wells, steam valves, chemical plants, aeroplane engines and vehicle engines.

5.3.3 THE INTERFACE

A typical composite contains about 60% by volume of fibres with a diameter of, say, $6\,\mu m$. A small piece of such a composite, say $10\,cm^3$ or $15\,g$, contains about $4\,m^2$ of interface area between the matrix and the reinforcing fibres. In a brittle matrix, strong bonding between matrix and fibres may not be advantageous, but with a tough matrix a strong interface will give optimum strength.

The interface between matrix and fibres is extensive.

The forces which operate at the interface are mechanical or chemical or a combination of both. Some examples are given of fibre-reinforced materials and also of particle-reinforced materials which you will meet later in the chapter.

FIBRE-REINFORCED MATERIALS

● In reinforced concrete, the bonding between the steel rods and the concrete matrix is mechanical [see § 5.5].

● In glass-fibre-reinforced thermosetting resins, the polyesters most commonly used bond to fibres by van der Waals forces.

● In fibre-reinforced thermoplastic resins, phenolic resins have high viscosity at the moulding temperature, which makes fibre impregnation difficult. Bonding at the interface is not strong.

PARTICLE-REINFORCED MATERIALS

● In particle-reinforced polymers [§ 5.6.1] friction and van der Waals forces operate between the particles and the matrix.

● In particle-reinforced metals [§ 5.6.2], chemical bonds form between the particles of e.g. aluminium oxide and the matrix, e.g. aluminium.

● Cermets have particles of ceramic dispersed in a metal matrix. The ceramic catalyses the corrosion of the surface of the metal matrix, and metal oxide grows into the crystal structure of the ceramic. Thus strong chemical bonds form at the interface [see § 5.12.1].

● A number of composites which are used in dentistry are described in § 6.8.2. In glass poly(alkenoate) cements, chemical reactions between the polymer and the glass matrix form strong chemical bonds at the interface. In polymer–ceramic composites, the ceramic particles are coated with a bonding agent which forms chemical bonds to the ceramic and to the polymer matrix. Bioglass, a ceramic, forms a composite material with bone by allowing regenerating bone to grow into it.

5.3.4 METHODS OF ASSEMBLING COMPOSITES

1. The fibres are assembled in a mould; then the matrix is introduced. This method needs highly fluid matrices, such as a vaporised matrix or a liquid metal or a monomer which can be injected as a liquid and subsequently polymerised.

2. The fibres are assembled as a woven fabric or a random mat and then impregnated with partially polymerised matrix, e.g. an epoxy resin or polyester resin. The intermediate material which is formed is then shaped into the required form, such as a surfboard or a canoe, and 'cured' (heat-treated) to make the matrix polymerise into its final form. Many people succeed in constructing their own boats and surfboards from glass-fibre-reinforced polyester in this way.

Methods of assembling composites are described.

3. The fibres are impregnated with a linear-chain polymer, e.g. poly(propene) or poly(etheretherketone). The composite can be rapidly shaped by one of the methods of moulding plastics shown in § 4.15.

5.4 PROPERTIES AND APPLICATIONS

Stiffness and low density are the outstanding characteristics of composites. These features are especially important in the aircraft industry. Fibre-reinforced plastics are used in laminated panels, wing spars, fuel tanks and bulkheads of aeroplanes and in helicopter rotor blades. The Boeing Chinook helicopter contains 20% by mass of such composites. A structure which contains highly aligned fibres will be stiff. Designers can therefore construct a structure of the stiffness required to meet the loads which it must bear. Some structures are designed to be very stiff in one direction but flexible in another (e.g. Kestrel bicycle, § 5.12.2).

Composites are valued as low-density, stiff materials . . .

FIGURE 5.4A
Chinook Helicopter

... can be designed to have a zero coefficient of thermal expansion ... ... and to survive in hostile environments.

Reinforcing fibres which are highly oriented can have negative thermal expansivity (contract on heating). It is therefore possible to design a structure with zero thermal expansion. This is important in the field of precision instruments. It is also essential in structures which must operate over a wide temperature range, e.g. satellites which encounter rapid changes in temperature between full sunlight and total shade.

Composites with a polymer matrix may cost more than metals ... ... but they are easier to fabricate ... ... and resist corrosion better.

Composite materials with polymer matrices are usually more expensive than metals. The cost of manufacturing an article may well be less, however, because it is possible to cast the polymer in a complex shape and therefore reduce the number of parts involved in building a structure. The Kestrel bicycle has a frame which is cast in one piece [see § 5.12.2]. Corrosion resistance is another factor in favour of composites because it gives a polymer matrix structure a longer lifetime.

CHECKPOINT 5.4

1. (*a*) Explain why a fibre has a higher tensile strength than a thicker piece of the same material.

(*b*) Explain why incorporating fibres into a plastic makes it stronger.

(*c*) Comment on the relative usefulness of long and short fibres.

2. Give three examples of types of materials which are used as matrices in composites and three examples of reinforcing fibres.

3. Describe briefly how each of these composites can be assembled:

(*a*) a fibre-reinforced epoxy resin

(*b*) a fibre-reinforced poly(propene)

(*c*) a fibre-reinforced metal

5.5 REINFORCED CONCRETE

Reinforced concrete is the material which is used in large structures, such as high-rise buildings, bridges and oil platforms. Concrete [see § 3.6] is weak in tension, strong in compression and brittle. When it is reinforced with steel wires or bars, which are strong in tension, the combination is a tough, relatively cheap material. A structural beam made of plain concrete would need to be very large and cumbersome to achieve the required strength [see Figure 5.5A]. For reinforcement, bars of cold-worked mild steel are used, often manufactured in spiral form to enable the concrete to adhere to them. The mechanical keying is improved by surface irregularities and by a thin layer of rust on the steel and by the shrinkage of concrete as it hardens round the steel rods. The purpose of the reinforcement is to make the strains in concrete and steel equal [see Figure 5.5A].

FIGURE 5.5A
Reinforced Concrete

(a) When a concrete beam is loaded, the top of the beam is subjected to compression

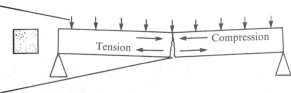

A plain concrete beam can fail when small cracks form in the edge which is under tension and act as stress raisers

(b) A plain concrete beam which is strong enough is too bulky

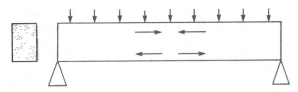

(c) A steel reinforcement bar is near the bottom of the bar. The ends are shaped so that they are gripped by the rigid concrete

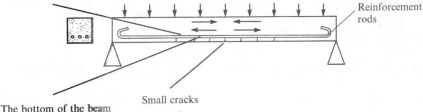

Reinforcement rods

The bottom of the beam is subjected to tension

Small cracks

The values of tensile (Young's) modulus are: steel $200\,\text{GN}\,\text{m}^{-2}$ and concrete $20\,\text{GN}\,\text{m}^{-2}$. Since [see §1.6.1]

$$\text{Tensile modulus} = \text{Stress/Strain,}$$

$$E = \sigma/\varepsilon$$

Putting strain in steel = strain in concrete,

$$\sigma_{\text{steel}}/E_{\text{steel}} = \sigma_{\text{concrete}}/E_{\text{concrete}}$$

$$\sigma_{\text{steel}}/\sigma_{\text{concrete}} = E_{\text{steel}}/E_{\text{concrete}} = 200\,\text{GN}\,\text{m}^{-2}/20\,\text{GN}\,\text{m}^{-2} = 10$$

Reinforced concrete is widely used in construction. Steel bars reinforce concrete because they have a higher tensile modulus.

The stress which causes concrete to fracture is $2.5\,\text{MN}\,\text{m}^{-2}$. When concrete fractures, the stress in the steel is $25\,\text{MN}\,\text{m}^{-2}$, which is well below its maximum stress. A concrete beam can carry a compressive load in the upper part of the beam while the steel rod carried the tensile load in the bottom of the beam. Cracks appear in the concrete due to tension, but the beam does not break. In time the cracks can allow water to penetrate and corrode the steel.

5.5.1   STRENGTHENING CONCRETE

The reason why unreinforced concrete has low tensile strength is that it contains microscopic pores. One method of reducing the porosity and increasing the tensile strength is to drive the air out of the cement powder before mixing it with water. This can be done by vibrating the powder. Another method is to add to the cement–water mixture a material, such as sulphur or a water-soluble polymer, which will fill the spaces between cement particles. Other filler materials include glass fibre and particles and fibres of silicon carbide and aluminium oxide.

Reducing the porosity of concrete strengthens it. This can be done by . . .
. . . driving out air
. . . adding a filler.

5.6   PARTICLE-REINFORCED MATERIALS

5.6.1   REINFORCED POLYMERS

Many polymeric materials incorporate fillers such as glass beads, silica and rubber particles. The toughness of some polymers is increased by incorporating tiny rubber particles in the polymer matrix. Poly(phenylethene) is toughened in this way by poly(butadiene) to give high-impact poly(phenylethene), HIPS (standing for high-impact poly(styrene)). HIPS is used for refrigerator trays and boxes and other household items and in toys. The rubber particles block the transmission of cracks [see § 1.10.1] and, since they deform easily, absorb energy. In the copolymer propenenitrile–butadiene–phenylethene, ABS (standing for acrylonitrile–butadiene–styrene), the butadiene rubber particles account for about 30% of the volume of the composite. ABS is used for protective helmets, battery cases, radio cabinets, water pumps and vehicle radiator grills.

Particles of glass, silica, rubber, etc. can be incorporated in polymers. Rubber particles block the transmission of cracks . . .
. . . and also lower the tensile modulus . . .
. . . and the tensile strength.
Carbon black is used as a filler in vulcanised rubber.

Carbon black is widely used as a filler in vulcanised rubber to enhance strength, stiffness, hardness, wear resistance and heat resistance of the rubber.

5.6.2   DISPERSION-HARDENED MATERIALS

Some composite materials are developed for tensile strength, rather than for hardness. One method of obtaining a strong composite is to disperse very small particles of aluminium oxide through a suitable matrix. One such matrix is aluminium itself. The composite can be made by taking powdered aluminium and grinding it in the presence of oxygen under pressure. Much of the surface film of aluminium oxide disintegrates to form a very fine powder intermingled with aluminium. The mixture is heated to give a homogeneous aluminium matrix containing about 6% of aluminium oxide particles. It is better than aluminium in tensile strength, especially at high temperature. The aluminium oxide present has a **dispersion-hardening** effect.

Aluminium oxide particles dispersed through a matrix increase its tensile strength.

Aluminium oxide particles are used for dispersion-hardening of other materials including silver and nickel. As mentioned in § 5.5.1 aluminium oxide particles and silicon carbide particles are used to improve the tensile strength of concrete.

5.6.3   LUBRICANT COMPOSITES

A number of modern mechanisms, particularly in aerospace engineering, require solid lubricants because liquid lubricants are unstable at high temperature. A solution to the problem is to incorporate a lubricant into the surface of the working parts. Composite materials incorporating graphite, molybdenum sulphide and compounds

Some composites include a lubricant with a similar layer structure may be employed and also poly(tetrafluoroethene), PTFE [see §4.11]. These substances lubricate as graphite does, by virtue of the weak van der Waals forces between layers [*ALC*, §6.6]. Examples are bronze bearings impregnated with PTFE or graphite and nickel, iron and other metals impregnated with molybdenum disulphide.

CHECKPOINT 5.6

1. (*a*) Explain how reinforcing a concrete beam with steel bars gives the beam a higher tensile strength.

(*b*) State two other ways in which the tensile strength of concrete can be increased.

2. State briefly how the mechanical properties of glass-fibre-reinforced polyester differ from those of (i) polyester, (ii) glass fibre.
Give examples of uses of the material and point out how its properties equip it for these uses.

3. The initials HIPS stand for high-impact poly(styrene).

(*a*) What is the systematic name for styrene, $C_6H_5CH{=}CH_2$?

(*b*) What reinforcing material is added to make the polymer 'high-impact'?

(*c*) How does this reinforcing material work?

4. The initials ABS stand for acrylonitrile–butadiene–styrene.

(*a*) What is the systematic name for acrylonitrile, $CH_2{=}CHCN$?

(*b*) What is the formula of buta-1,3-diene?

(*c*) What effect does the incorporation of butadiene have on the mechanical properties of the polymer?

5. Epoxy resins are thermosetting plastics. (*a*) Why does this make them liable to crack? (*b*) How does the incorporation of carbon fibres stop cracks spreading? (Refer to §1.10.1 if necessary.) (*c*) Give two examples of the use of this material in applications where cracking must be avoided.

5.7 FOAMS

In foams the matrix is a gas. Foams are a type of composite in which the component bound by the matrix is not a solid, but bubbles of gas. Such foams are used as cushioning in furniture, packaging and padding, for thermal insulation, for buoyancy and as filling in laminates [see §4.14].

5.8 LAMINATES

Many materials are stronger in one direction than another.
In a laminated material, layers of the material are arranged so as to make the laminate stronger than a block of the material.
In wood, the tensile and compressive strengths are very much greater along the grain than in a direction at right angles to the grain. The shearing strength along the grain is much lower than the shearing strength across the grain. The grain directions are the directions of the cellulose fibres in the wood. A material which has different properties in different directions is **anisotropic.** One way in which the **anisotropy** of wood can be overcome is by bonding thin layers together as plywood. Thin layers of wood are bonded together with a water-resistant glue or a thermosetting resin with the grains of successive layers at right angles to one another. Plywood is an example of a **laminated material** [see Figure 5.8A].

FIGURE 5.8A
Plywood

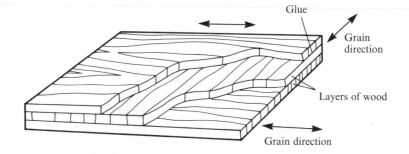

Metals also can be laminated. Aluminium–copper alloy is coated with aluminium to improve its corrosion-resistance. Galvanised steel is steel that has been plated with zinc to inhibit rusting. Food containers are made from steel plated with tin to improve resistance to corrosion.

5.8.1   SANDWICH STRUCTURES

Sandwich structures are used when a combination of high stiffness and low mass are required. They are composed of two skins of high strength with a low-density core.

Corrugated cardboard is a laminated material [see Figure 5.8B]. The sandwiching of a layer of corrugated paper between sheets of paper makes a structure which is stiffer in the direction parallel to the corrugations than paper alone.

FIGURE 5.8B
Corrugated Cardboard

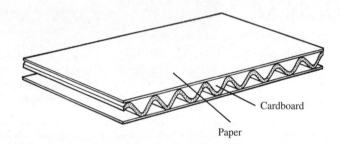

Cardboard

Paper

Sandwich structures can be stiff and of low density.

A sandwich structure is stiff and of low density. The skin materials include sheet metals, plywood, plastics, including fibre-reinforced plastics, concrete and plaster-board. The core may be paper, plastic foam, chip board, glass-fibre-reinforced resin or a honeycomb structure of thin sheet metal, e.g. aluminium, titanium or steel.

A sandwich structure may be used as a structural element in aircraft. Aluminium is often used both for the honeycomb and for the sheets [see Figure 5.8C].

FIGURE 5.8C
A Honeycomb Structure which is Part of an Aircraft Wing

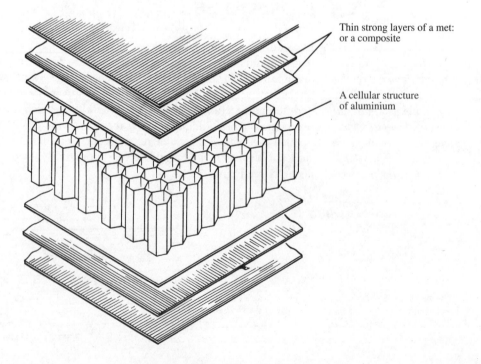

Thin strong layers of a met:
or a composite

A cellular structure
of aluminium

5.9 CELLULAR MATERIALS

Wood is a material of low density – about $0.5\,\mathrm{kg\,dm^{-3}}$. Yet a tree consisting of a column of wood can rise to a height of 30 m or more and support many branches and thousands of leaves, sometimes with a load of snow and ice, while resisting high wind forces. The high specific tensile strength (tensile strength/density) of wood results from its internal structure, which is cellular [see Figure 5.9A].

FIGURE 5.9A
Wood Cells

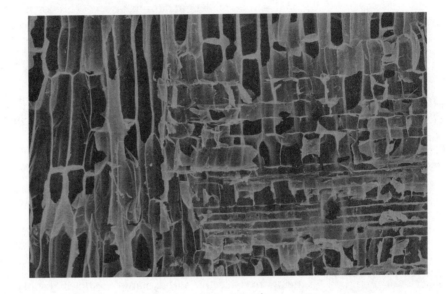

Cellular materials are strong, e.g. cork, wood, honeycomb structures.

In cork, the cell walls, which are composed of cellulose, and other structures make up 10% of the volume of the cell. The remainder is air. This structure is responsible for the ability of cork to insulate against heat and sound.

Synthetic materials imitate the cellular structure of sapwood and cork. An example is the paper honeycomb structures, which are used for packaging and in internal doors.

Aluminium honeycomb panels have become a feature of aircraft body construction [see Figure 5.8C] and have spread to other areas – boats, racing cars, skis and other sports products [see § 5.10.2]. Aluminium alloy foil 0.04 and 0.08 mm thick is converted into hexagonal cells 5 cm across. Outer skins of aluminium sheet or plastic are secured to the honeycomb by adhesive. The honeycomb panels have excellent compressive strength and shear strength and combine lightness with stiffness.

CHECKPOINT 5.9

1. (a) Explain why glueing together thin sheets of wood can produce a stronger material than the same thickness in one piece.

(b) Give two examples of laminates.

2. The value of a reinforcing material is not dependent only on its tensile modulus and its tensile strength. What else needs to be taken into account?

3. What effect does a fibre reinforcement have on (a) the fatigue failure of a metal and (b) the creep resistance of a metal?

4. In what type of matrix will the inclusion of fibres result in (a) a reduction of impact strength, (b) an increase in impact strength?

5.10 SOME EXAMPLES OF THE USE OF COMPOSITE MATERIALS

5.10.1 CERMETS

Cermets are composed of ceramic and metallic components.
They are more elastic and less brittle than ceramics . . .
. . . and harder than metals . . .
. . . finding uses in metal-cutting tools and in vehicle engine components.
Bonds form between the metal and the ceramic.

Cermets are a set of composite materials. They are made by compacting a mixture of powdered ceramic and metallic components and heating the mixture. The ceramic acts as a catalyst which assists the metal to corrode to form its oxide. As the metal oxide is formed, crystals of the oxide grow into the crystal structure of the ceramic. The bond is strong and permanent and forms rapidly. This bonding can occur between all metals and most ceramics. The strongest bonds are formed between noble metals, transition metals and the oxides of aluminium, magnesium, silicon, titanium and zirconium.

Cermets include **cemented carbides**. These consist of hard particles of tungsten carbide or titanium carbide in a matrix of cobalt.

The physical properties of cermets are found neither in ceramics nor in metals. They are elastic, hard and brittle, but they have low impact strength and low resistance to thermal shock. They are used in linings for brakes and clutches, in non-lubricating bearings, for the tips of metal-cutting tools and in dies for drawing wires. Other applications are gold-coated ceramic wafers for semiconductor chips, zirconium oxide-lined steel for corrosion-resistant uses and ceramic-capped dental crowns [see §6.8.2].

5.10.2 BICYCLE WHEELS AND FRAMES

Racing bicycles have disc wheels . . .
. . . made of two carbon-fibre-reinforced epoxy resin sheets . . .
. . . with a polystyrene foam or a honeycomb structure sandwiched between them.
Racing bicycle frames have been made of fibre-reinforced epoxy resins.

Bicycle wheels have a hub in the centre with spokes radiating out to the rim. In an attempt to reduce the aerodynamic resistance of spokes, disc wheels have been adopted by racing cyclists. Carbon-fibre-reinforced composites are used in the manufacture of disc wheels. The disc wheel is composed of two carbon-epoxy sheets with a polystyrene foam or a honeycomb sandwiched between them. The carbon–epoxy layers are fastened to an aluminium hub and to a carbon–epoxy rim.

Bicycle frames have also been made of composites. The Kestrel bicycle has an epoxy resin matrix strengthened with carbon fibres and poly(ethene) fibres. The frame is moulded in one piece. Because of the care that has to go into the alignment of the fibres, the manufacture is not automated. A Kestrel cycle is very light: the frame weighs about 1.6 kg and is twice as strong as a steel frame. It is built to be stiff in the

---

The carbon monocoque

Chris Boardman rode a cycle with a one-piece frame of carbon-fibre-reinforced epoxy resin when he gained the Olympic Gold Medal in Barcelona in 1992. In the Tour de France of 1994 he recorded the fastest ever average speed in a time trial and wore the leader's yellow jersey for three days. The Lotus Sport 100 carbon monocoque is a production model which has been developed from the revolutionary new design and which can be yours for £6000.

FIGURE 5.10A
Lotus Sport 100
Carbon Monocoque

horizontal dimension and to have some give in the vertical dimension to achieve shock absorption and a comfortable ride. The cost of materials and the complicated manufacture make the cycle expensive (£1200 at 1990 prices).

You can see how composites compare with other materials in cycle manufacture in §6.7.6.

5.10.3 WINDSURF BOARD

Figure 5.10B shows the composition of a windsurf board. You can see the number of components that are used to give strength coupled with lightness. Note also the use of honeycomb structures for the combination of high strength and low density.

FIGURE 5.10B
A Cross-Section of a
Windsurf Board

Honeycomb structures are composite materials. They are used in the construction of windsurf boards.

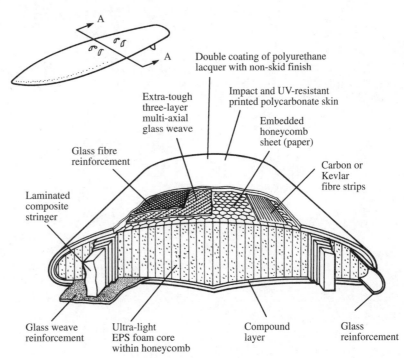

Section on A-A

═══════════ **QUESTIONS ON CHAPTER 5** ═══════════

1. Explain the term 'composite'.

2. Give examples of composites involving (*a*) metals, (*b*) plastics.

3. (*a*) Why do glass fibres have different properties from plate glass?

(*b*) What advantage does glass fibre-reinforced polyester have for building boats (i) over wood and (ii) over metal?

(*c*) Describe how the mechanical properties of a fibre composite depend on the form of the fibres and their orientation.

(*c*) Give one disadvantage of GRP compared with steel for use in vehicle bodies.

4. For each material list the uses to which it can be put. Each material may have more than one use.

Material	Use
1. Alloy	A Canoe
2. Ceramic	B Bridge
3. Glass-fibre reinforced polyester	C Electric light wall switch
4. Thermosoftening plastic	D Rifle
5. Thermosetting plastic	E Tennis racquet
6. Synthetic fibre	F Packaging
7. Carbon fibre-reinforced epoxy resin	G Carrier bag
8. Reinforced concrete	H Ovenware
9. Paper honeycomb	I Clothing

5. (*a*) What is a 'cermet'?

(*b*) How do cermets differ from (i) ceramics, (ii) metals?

(*c*) Give two examples of the uses of cermets.

6. Explain how particles of carbon black improve the tensile modulus of rubbers.

7. Explain why corrugated cardboard and plywood are stiff.

8. (*a*) Describe the essential characteristics of a composite material. (*b*) Explain the advantages of reinforced concrete over ordinary concrete.

9. The underside of a concrete bridge cracks, exposing steel reinforcing rods. What is the likely cause of the damage? How might the problem be avoided in the construction of future bridges?

10. Give two examples of the use of composites to replace metals in parts of bicycles.
Explain why composites are chosen, rather than metals, in the examples you have given.

6
REVIEW OF MATERIALS

6.1 CLASSIFICATION

One way of classifying materials is to divide them into four groups on the basis of their structure:

Group 1. Amorphous materials that have the atoms randomly arranged

Group 2. Crystalline materials that have atoms arranged in almost perfect order

Group 3. Composite materials that consist of two or more different materials

Group 4. Cellular materials that consist of stacks of hollow symmetrical cells

Materials can be classified on the basis of their structures . . .
. . . as amorphous . . .
. . . crystalline . . .
. . . composite . . .
. . . or cellular.
Each class includes metals, ceramics and polymers.

Each of these groups includes metals, ceramics and polymers in various proportions. The only ceramics in Group 1 are glass ceramics, and the only polymers in Group 2 are those with an exceptionally high degree of crystallinity. The manufacturer and the engineer have a huge number of materials to choose from. There are about 15 000 different plastics available, and the number is increasing. New metallic alloys are being made from combinations of different elements in different proportions. New ceramics are being produced and new heat treatments are being devised to change the properties of existing materials.

6.2 BOND TYPE AND PROPERTIES

You have studied materials of many types. It is now time to draw these materials together, to compare the properties of different solid materials, and to see how these properties make different materials useful for different purposes.

6.2.1 METALS

Metals [see Chapter 2]:

- are generally stiff, hard, tough and shiny
- change shape without breaking (are ductile and malleable)
- are mostly strong in tension (stretching) and compression (loading)
- are good thermal and electrical conductors
- are in many cases corroded by water and acids.

The properties of metals result from the nature of the metallic bond [see § 2.3] and the crystalline structure of metals [see § 2.5]. The metallic bond is non-directed in space. Metals are strong in tension and in compression because the bonds between the atoms are strong. Metals are giant structures in which the delocalised electrons are free to move, and these mobile electrons allow metals to conduct heat and electricity. Metals can change shape without breaking because the metallic bond allows layers of atoms to slip over one another when the metal is under stress [see § 2.7.1]. The crystalline structures of metals contain dislocations [see § 2.7.2] which will move under stress and allow the metal to change shape. The movement of dislocations provides a much easier method of deformation than slip. Metals are weaker in compression than glass: they can withstand smaller loads without changing shape than can glass.

The properties of metals . . .
. . . including strength and
conductivity . . .
. . . are explained in terms
of the nature of the metallic
bond . . .
. . . and the crystalline
structure of metals . . .
. . . including dislocations.

6.2.2 CERAMICS

Ceramics [see Chapter 3]:

- are very hard and brittle
- are strong in compression but weak in tension
- are electrical insulators
- have very high melting temperatures
- are chemically unreactive.

The properties of
ceramics . . .
. . . including hardness,
brittleness and high
melting temperature . . .
. . . are explained by
reference to the crystalline
structure . . .
. . . which is maintained by
ionic and covalent bonds.

Ceramics [see § 3.1] are crystalline compounds of metallic and non-metallic elements. In most ceramics, the atoms which form the framework of the structure are linked by covalent bonds which are directed in space. The structure may incorporate metal cations which are linked by electrovalent bonds.

The structure of a ceramic is therefore more rigid and less flexible than that of a metal. This structure makes ceramics harder than metals and also more brittle. Ceramics have lower densities and higher melting temperatures than metals.

Lacking free electrons, ceramics are poor electrical and thermal conductors.

6.2.3 GLASSES

Glasses [see Chapter 3]:

- have lower melting temperatures and glass transition temperatures than other ceramics
- are generally transparent

Glasses are ceramics below
their glass transition
temperatures.
Most consist of covalently
bonded silicate structures
associated with metal
cations.

Glasses [see § 3.7] are a subset of ceramics. The structure of most glasses is a covalently bonded framework of silicate tetrahedra with metal cations linked by electrovalent bonds [see Figure 3.5F].

A plane of regularly spaced atoms in a crystal structure forms a surface which can reflect light. Glass is transparent because it is non-crystalline, so light can pass through without meeting any reflecting surfaces.

Glasses are transparent
because they are
amorphous . . .
. . . and brittle because of
surface defects.

Glass should be a strong material because the bonds between atoms are strong and there are no dislocations. It is able to withstand mechanical loading without breaking. However, glass is brittle because of the rigid bonds and shatters easily. The brittleness is increased hugely by surface defects or scratches [see § 1.10 and § 3.10]. Glass breaks easily if a scratch is first made on its surface. Cracks find it harder to grow in metals because of the movement of dislocations.

Glass fibres can be bent without cracking because they are free of surface scratches.

6.2.4 PLASTICS

The properties of plastics, including strength and plasticity, are related to their structure ...
... a tangled mass of very long polymer molecules ...
... with forces of attraction between molecules ...
... which may be weak or strong.
Some plastics have an amorphous structure ...
... some are largely crystalline ...
... and some have a mixture of amorphous and crystalline regions.
The wide variations in structure ...
... correspond to a wide range of properties.

Plastics [see Chapter 4]:

- are usually strong in relation to their density
- are mainly (but not all) soft, flexible and not very elastic
- soften easily when heated and melt or burn
- are thermal and electrical insulators.

The bonding in plastics is covalent. Plastics consist of a tangled mass of very long polymer molecules. In thermosoftening plastics, the bonds between chains are weak van der Waals forces; in thermosetting plastics covalent bonds form cross-links between chains.

Most polymers are poor electrical and thermal conductors because they lack free electrons.

Articles made of plastic can be coloured by mixing pigment with the plastic during manufacture. They have the advantage over painted articles that the colour extends all through the article and cannot be chipped off. Articles made from metals and alloys are mainly grey in colour until paint is applied. The exceptions are copper and gold and their alloys and anodised aluminium (aluminium which has been made the anode of an electrolytic cell) which can take up dyes.

Synthetic fibres: For the structure of synthetic fibres see § 6.3.2.

6.2.5 CONCRETE AND MOST TYPES OF ROCK

Concrete and most types of rock are:

- strong in compression (loading)
- weak in tension (stretching).

Concrete has a silicate framework associated with metal cations.

Concrete consists of a covalently bonded silicate framework associated with metal cations, including calcium, aluminium, iron (II) and magnesium ions. Reinforced concrete is described in § 5.5.

6.2.6 COMPOSITE MATERIALS

Some materials are not homogeneous. They include composite materials, e.g. reinforced concrete, and cellular materials, e.g. cork [see § 5.9]. In a composite material, the properties of the components combine to give a material which is more useful for a particular purpose than the individual components. Examples are:

REINFORCED CONCRETE

Concrete [see § 3.6] is weak in tension, strong in compression and brittle. When it is reinforced with steel wires or bars, which are strong in tension, the combination is a tough, relatively cheap material [see § 5.5].

FIBRE-REINFORCED PLASTICS

Composite materials combine the properties of their components, e.g. steel-reinforced concrete e.g. fibre-reinforced plastics.

Epoxy resin and polyester resin are thermosetting plastics which can be reinforced by fibres to form composite materials. These composites can be used for some applications which were traditionally filled by metals. Examples of fibre-reinforced plastics are:

- glass-fibre-reinforced polyester [see § 5.3.2]
- carbon-fibre-reinforced epoxy resin [see § 5.3.2]

LAMINATED PLASTICS

See § 4.12.6.

6.2.7   BONDING SUMMARISED

Metals: the metallic bond

Polymers: covalent bonds and intermolecular forces (van der Waals forces, dipole–dipole interactions and electrostatic attractions) which hold chains together

Ceramics: both covalent and electrovalent bonding, e.g. covalent bonding in silicate chains and electrovalent bonding between silicate chains and metal cations

The different types of chemical bond are summarised.

The complexity of many ceramic structures and polymer structures makes crystallisation difficult, so these materials frequently form glasses.

6.3   STRUCTURES

6.3.1   CRYSTALLINE AND AMORPHOUS MATERIALS

Crystalline materials have perfectly ordered arrangements of atoms. Amorphous materials have random arrangements of atoms.

In **crystalline** materials the atoms are stacked up in perfectly ordered symmetrical arrangements. In **amorphous** materials, combinations of atoms have little or no order or symmetry.

METALS

Metals have a crystalline structure in the solid state ... ... except for glassy metals.

In metals the distinction between crystalline and amorphous is clear: liquid metals have a completely random arrangement, and solid metals can be perfectly crystalline. The crystal structures of metals are illustrated in § 2.5. When a liquid metal is cooled extremely rapidly (one million degrees per second) the random arrangement of atoms that occurs in the liquid is 'frozen' in the solid. Such metals are described as 'glassy metals' because the structure has some similarity to glass. The structure of glass is not exactly the same as that of an amorphous metal: there has to be some degree of order to maintain the silicon/oxygen ratio.

PLASTICS

Ceramics are crystalline above the glass transition temperature ... ... and glasses below the glass transition temperature.

The structure of some plastics is described as **amorphous** (shapeless) because the polymer chains take up a completely random arrangement. Many synthetic polymers have a structure which contains a mixture of amorphous and crystalline material, depending on how the individual polymer chains are arranged with respect to their neighbours. In some regions the polymer chains may be packed together in a regular **crystalline** manner; for example, low-density poly(ethene) has a mixture of crystalline regions and amorphous regions [see Figure 4.9A]. By contrast, in high-density poly(ethene), the chains pack closely together with a high degree of crystallinity [see Figure 4.9B]. This is why high-density poly(ethene) is stronger than low-density poly(ethene) and is not as easily deformed by heat. There is wide variation in the degree of crystallinity, chain length, chain branching, cross-linking within molecules and between molecules. The variations in structure are accompanied by a wide variety of properties, including those of Perspex, PVC and nylon.

Many synthetic polymers have a mixture of amorphous and crystalline regions.

FIBRES

Fibres are made by drawing (stretching) plastics by melt spinning, dry spinning or wet spinning [see Figure 4.16A]. As the fibre solidifies, it is stretched to align its molecules along the length of the fibre. This is why fibres have great tensile strength along the length of the fibre.

6.3.2 DEFORMATION

Crystalline materials can be deformed without fracturing . . .
. . . due to the movement of dislocations.

Crystalline materials can undergo plastic deformation. In metals, this can take place by means of slip [§ 2.7.1] but is made much easier by the presence of dislocations [see § 2.7.2], which allow atoms in one crystal plane to move relative to an adjacent plane. All crystalline materials contains millions of dislocations per cubic millimetre.

In ceramics the movement of dislocations is difficult . . .
. . . and ceramics are brittle. Metals are ductile and tough.
Ceramics have poor ductility and low toughness.

Ceramics have a more complicated arrangement of atoms than that in metals, which makes dislocation movements very complicated. Also bonds between atoms in ceramics are more directed than in metals, making plastic deformation a difficult process. Ceramics are usually brittle. Metals and ceramics represent two extremes of crystalline behaviour from good ductility and high toughness (metals) to poor ductility and low toughness (ceramics). However, most ceramics are harder than most metals, and most have higher melting temperatures because of the oriented covalent bonding.

Polymers are deformed in the same way.
Some can be deformed by straightening the polymer chains . . .
. . . while others are brittle.

Polymers cannot be deformed in the same way as metals because they consist of rigid chains of atoms. Some polymers can be deformed by straightening out their component chains, e.g. rubbers, polyesters, polyamides and polycarbonates. Brittle polymers include Perspex and polystyrene. If you break a plastic ruler, made of poly(methyl 2-methylpropenoate), about 90% of the fracturing is due to individual polymer chains being pulled out of the material, rather than the breaking of covalent bonds. In other polymers, the molecular chains are more densely packed and more entangled, and the plastics are more difficult to break. It is possible to alter the shape of some plastics by straightening out the polymer chains. Examples are rubbers, polyesters, e.g. Terylene, polyamides, e.g. nylon, and polycarbonates.

Glass is brittle because of its rigid, directed covalent bonds and non-crystalline structure.

6.4 STRUCTURAL MATERIALS

Considered as structural materials . . .
. . . polymers are too easily deformed . . .

Consider metals, ceramics and polymers as structural engineering materials. **Metals** have high tensile and compressive strength. They are ductile: they can change shape without fracturing. Metals can change shape through the movement of dislocations; this can only happen in crystalline materials. Their strength and ductility enable them to be used for structures which have to withstand stress, e.g. bridges, ships, motor vehicles, aircraft and machinery.

There are some limitations on the usefulness of metals. Pure metals are made stronger by alloying and by heat treatment, but as they become stronger their ductility is reduced. Metals and alloys soften at high temperatures, and every alloy has a temperature above which it should not be used. Most engineering metals are too dense to be used in aircraft manufacture. The exceptions are aluminium, titanium, magnesium and beryllium.

. . . metals are strong in compression and in tension and are ductile . . .
. . . polymers cannot withstand stress . . .
. . . ceramics are strong in compression but weak in tension and brittle . . .
. . . harder than metals and with higher melting temperatures . . .

Polymers have a low modulus of elasticity [see § 1.6] and often undergo larger plastic deformation on loading than do metals and ceramics. They cannot be used in the construction of structures which have to withstand stress.

Ceramics, both crystalline and glassy, tend to be brittle. Traditional ceramics are weak as well as brittle. Many modern ceramics, e.g. alumina, are brittle but also strong in compression, that is, they can bear a load. They can be used for applications where they have to withstand compression but not where they are subject to tension. They score over metals for tasks in which hardness at high temperature is required.

... while composites may be designed to combine compressive strength and tensile strength.

Composite materials are designed to combine useful properties, e.g. reinforced concrete combines the compressive strength of concrete with the tensile strength of steel.

Temperature Figure 6.4A shows the temperatures up to which different materials can be used.

FIGURE 6.4A The Temperatures up to which Materials Can Be Used

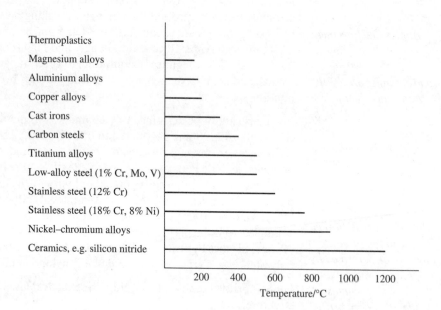

The temperature restriction on the use of materials is considered.

CHECKPOINT 6.4

1. Explain why it is possible to bend a sheet of poly(chloroethene) (PVC) more easily than (*a*) a sheet of metal, (*b*) a sheet of glass, (*c*) a sheet of Perspex.

2. Explain why a piece of ceramic is shattered by a hammer blow but a piece of metal merely bends.

3. (*a*) Why do glass fibres have different properties from plate glass?

(*b*) What advantage does glass-fibre-reinforced polyester have for building boats (i) over wood and (ii) over metal?

(*c*) Give one disadvantage of glass-fibre-reinforced polyester, GRP, compared with steel for use in vehicle bodies.

6.5 ENVIRONMENTAL STABILITY OF MATERIALS

6.5.1 METALS

Metals are corroded in air ...
... with exceptions, e.g. gold, aluminium and others.

The majority of metals are corroded in air. There are exceptions: metals low in the electrochemical series, e.g. gold and platinum. There are metals which corrode rapidly to acquire a coating of the oxide and then corrode no further, e.g. aluminium and chromium. For other metals, the rate of corrosion depends on a number of factors. These include the presence of water, air, electrolytes and other metals which can form a galvanic cell [see § 2.13].

6.5.2 POLYMERS

Polymers can be used up to 60–120 °C, depending on the identity of the polymer. The tensile strength and hardness of thermoplastics decrease with temperature. They show creep even at room temperature, and the tendency to creep increases with rising temperature.

Polymers soften and weaken at much lower temperatures than do metals and ceramics. Some polymers are resistant to chemical attack . . .

Some polymers are very resistant to chemical attack. Others may stain, crack, soften, swell or dissolve. Nylon and other polyamides are resistant to attack by weak acids and alkalis and by organic solvents but are attacked by strong acids and alkalis. Poly(phenylethene) can withstand attack by acids and alkalis but is attacked by organic solvents. Polymers in general are resistant to water.

Polymers are generally affected by exposure to the atmosphere and sunlight. Most thermoplastics are affected by ultraviolet light. Ultraviolet light can cause the breaking of bonds in the molecular chains and result in surface cracking. Many plastics incorporate a substance which can absorb ultraviolet light without damage.

. . . while others are hydrolysed . . .
. . . and some dissolve in organic solvents. Polymers are weakened by exposure to sunlight and oxygen.

Natural rubber is resistant to acids and alkalis but has poor resistance to petroleum products. It deteriorates rapidly in sunlight. Some synthetic rubbers, e.g. neoprene, are more resistant to petroleum products and sunlight. Ozone causes cracking in natural rubber and many synthetic rubbers. Neoprene has a high resistance to ozone.

FIRE HAZARDS

Many polymers are flammable . . .
. . . and some give toxic combustion products.

Many polymers burn easily. Some give carbon monoxide as a combustion product, and nitrogen-containing polymers, e.g. poly(urethane), produce hydrogen cyanide. Poly(urethane) foams, which were widely used for cushion-filling, burn very readily, producing dense smoke and toxic gases. Manufacturers have now discontinued using poly(urethane) for domestic furniture.

6.5.3 CERAMICS

Ceramics are relatively stable when exposed to the atmosphere. When sulphur dioxide is present in the atmosphere, it will cause deterioration, as when stone and brick are damaged by exposure to an atmosphere containing sulphur dioxide.

Ceramics are more stable than metals and polymers to exposure to the atmosphere . . .
. . . but some are attacked by acids.

Water is absorbed in the pores of ceramics, and if the water freezes and expands damage will occur. The low thermal conductivity of ceramics can result in large thermal gradients being set up, causing stress. The surface may flake as a result. This is a factor which has to be considered in the choice of ceramics for use as furnace linings.

6.6 A COMPARISON

Property	Metals	Ceramics	Plastics
Density/$kg\,dm^{-3}$	2–16 (average 8)	2–17 (average 5)	1–2
Melting temperature/°C	Low (e.g. Sn 232) to high (e.g. W 3400)	High, up to 4000	Low, up to 300
Tensile strength/MPa	Up to 2500	Up to 400	Up to 120
Compressive strength/MPa	Up to 2500	Up to 5000	Up to 350
Tensile modulus/GPa	40–400	150–450	0.7–3.5
Hardness	Medium	High	Low
Machineability	Good	Poor	Good
Resistance to creep at high temperature	Poor	Excellent	Cannot be used at high temperature
Thermal expansivity	Medium to high	Low to medium	Very high
Thermal conductivity	Medium to high	Medium	Very low
Resistance to thermal shock	Good	Poor	Melt
Electrical properties	Conductors	Insulators	Insulators
Resistance to chemical attack	Low to medium	Excellent	Good in general
Resistance to oxidation at high temperature	Poor, with exceptions	Good, and many are excellent	Cannot be used at high temperature

The table gives a comparison of metals, ceramics and polymers.

TABLE 6.6A
A Comparison of the Properties of Metals, Ceramics and Plastics

6.7 SOME CASE STUDIES

6.7.1 CONTAINERS FOR CARBONATED DRINKS

The sale of beer and carbonated soft drinks is big business. The choice of containers comprises glass bottles, plastic bottles and metal cans. The properties which a container for a carbonated drink must have are:

● inertness: The container must not be corroded by the atmosphere and must not react chemically with the carbonated drink.

● non-toxicity: The material must be perfectly safe to use for a beverage.

● strength: The material must be able to stand up to handling when it is packed, in transport and in use. Glass bottles pose more of a threat of breakage. They are safe when handled properly and do not break in transport. It is when people throw their empty bottles aside that the danger of broken glass to people and animals becomes a serious threat.

● impermeability to gases: The drinks must not go flat even if containers are stored for some time. Some plastics fail on this count.

Carbonated drinks are sold in glass bottles, plastic bottles, coated steel cans and aluminium cans. The material must be inert, non-toxic, impermeable to gases, low-density and low-cost.

● low cost: The price is determined by the cost of the raw materials and the cost of the fuel and electricity used in manufacture. The possibility of reusing containers and recycling materials will affect the cost.

● low density: If a low-density material is used, the filled containers weigh less and cost less to transport.

Glass, metals and plastics all have their advantages.

GLASS

A glass bottle has the appeal of being transparent and allowing drinks manufacturers to tempt their customers with the sight of the contents: pale yellow lemonade, nut-brown ale, orangeade or mineral water. Many plastics are partially transparent but none has the same appeal as glass. Glass bottles could be reused, but there is no scheme for collecting and reusing bottles. They can be recycled, and the Bottle Bank scheme, which is widespread, reduces the cost of glass bottles. Glass bottles suffered a setback when metal cans with ring pulls were invented because a bottle-opener was needed to open many glass bottles. Manufacturers have since remedied the situation by making bottle tops which can be screwed off by hand.

Glass has the advantage of being transparent, low-cost and refillable or recyclable.

PLASTICS

The choice of plastics has to be restricted to **commodity plastics**, the plastics which are produced in large quantities at low cost. These include poly(ethene), poly(propene), poly(chloroethene) (PVC) and poly(ethene benzene-1,4-dicarboxylate) (PET). The cheapest of these is poly(ethene), but poly(ethene benzene-1,4-dicarboxylate) (PET) is the least permeable: it can contain a carbonated drink without allowing gas to escape or the atmosphere to enter. This property is related to its structure. Crystalline polymers are less permeable to gases than amorphous polymers because crystalline polymers have fewer spaces in the structure than amorphous polymers. In general, polymers with low glass transition temperatures have low barriers to the diffusion of gases at room temperature.

Plastic bottles have the advantage of being lightweight and difficult to break.
A plastic which is impermeable to gases must be chosen, e.g. PET. The recycling of PET is possible but there is no scheme in place for collecting used bottles.

An advantage of plastics over glass is the lower weight of the plastic bottle. Very thin plastic bottles can be made and strengthened by drawing. Transport costs are reduced when lighter containers are used. Plastics have an advantage over glass in being difficult to break.

Some steps have been taken to collect and recycle plastics, but little recycling of drinks bottles takes place. (You can read about an exception in the Box.)

Twenty-seven old bottles make a new sweater!

French chemists have devised a process for making sweaters from a fabric which is 30% wool and 70% recycled PVC. The PVC comes from mineral water bottles, of which 3.2 billion are used annually in France. It takes 27 bottles to make a sweater. The bottles are ground into fine chips, melted, purified and spun into yarn. The difficulties of recycling are getting rid of the blue dye used to colour the bottles, the bottle caps which are of a different plastic, the labels and the additives used to strengthen the plastic. The sweaters cost £75–£100.

METALS

Metal cans have the advantage of being unbreakable.
Both coated steel and aluminium cans are widely used.
Steel is cheaper, but aluminium cans are lighter and therefore cheaper to transport.

Coated steel (to protect against corrosion) and aluminium are the contenders. Steel has the advantage of being a cheaper material than aluminium. However, recycling aluminium cans is easier than recycling steel cans and to some extent reduces the cost of aluminium cans. Making a can from recycled aluminium takes 5% of the energy required to make a can starting from bauxite. Aluminium is a low-density metal and cans made from it are therefore lighter than steel cans so the cost of transporting them is lower.

The manufacture of beer and soft drink cans runs to 150 billion aluminium cans a year and swallows 40% of the world's aluminium production. Each can is made from a 5 cm diameter disc by a drawing process which strengthens the metal by aligning the grains and introducing more dislocations. The drawing gives the can enough strength to withstand handling inspite of its paper-thin walls. Comparative weights are: a glass bottle 230 g, a tinplate can 35 g and an aluminium can 20 g.

The collection and recycling of used aluminium cans is well established.

The ease of opening the can has proved to be an important factor,. In 1963 Ernie Fraze of the USA invented his easy-open end for carbonated drink cans. The design features a tab and a score line that enables the consumer to pull the tab, which tears round the score line and opens the can without the use of a can opener. To date about 2 million million of Fraze's ring-pull ends have been used. By far the best material for ring-pull cans is aluminium as it tears easily from the score line. In 1991 in the USA of 98 billion cans produced 97% used aluminium, while in Europe out of 21 billion cans 60% used steel and 40% used aluminium. A drawback to the original ring-pulls was that many people threw them away to litter the environment. They have now been designed to remain attached to the can.

In 1983 European producers of steel cans started to collaborate on the production of a steel easy-open can for carbonated drinks.

Both aluminium and steel cans are supplied with easy-to-open tabs which are popular with customers.
All four materials are extensively used.

They wanted a push-in tab which is easy to open and is not discarded as litter. The score is thinner in steel than in aluminium because steel is stronger.

The relative costs of materials used to make bottles are shown in Table 6.7A.

TABLE 6.7A
Relative Costs of Materials Used to Make Bottles

Construction material	Relative cost
Aluminium alloy circles for drawing	130
Mild steel, cold-drawn	22
PET	70
Glass	10

All four materials are extensively used: glass, poly(ethene benzene-1, 4-dicarboxylate), aluminium and coated steel.

6.7.2 DENTAL RESTORATIVE MATERIALS

Tooth enamel is composed mainly of hydroxyapatite, $Ca_5(PO_4)_3OH$. Bacteria in the plaque on teeth produce acids which can attack the enamel. When a tooth develops **caries** (decay), a dentist removes the decaying material and fills the cavity. The material chosen to fill the cavity may be a metal, an alloy, a ceramic or a composite material. The material must fulfil a number of requirements.

1. It must be malleable so that it can be moulded to fill the cavity completely.

2. The material must adhere well to the tooth.

3. The contraction on setting must be small so that no gap is formed between the filling and the tooth, allowing fluid to leak in.

Materials which are used for dental restoration ...
... must be malleable ...
... must adhere well to the tooth ...
... must be hard-wearing and strong ...
... must not be corroded by fluids in the mouth ...
... must be biocompatible.

4. The material must be able to withstand abrasion and the force of chewing.

5. It must not be corroded or discoloured by fluids in the mouth. The mouth contains a solution of salts at pH 7.4 and temperature 37 °C. These conditions promote electrochemical corrosion of metals and the breakdown of hydrolysable polymers. The materials used for filling cavities, making crowns, false teeth, etc. must be stable under these conditions.

6. The materials must be biocompatible. **Biocompatability** means that the foreign material and the natural tissue remain functioning together, without either significantly affecting the other. Originally the search was for materials that are completely inert, e.g. gold, alumina and porcelain. Now the aim is to find dental materials that bond with enamel and dentine so that the natural tissue accepts the foreign material instead of just ignoring it.

Methods used to restore teeth are illustrated in Figure 6.7A.

FIGURE 6.7A
Methods of Restoring Teeth

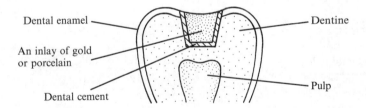

Dental enamel — Dentine
An inlay of gold or porcelain
Dental cement — Pulp

(a) An inlay or a crown of gold alloy or dental porcelain is made to fit the cavity. It is placed in the tooth and then cemented to the sound tissue.

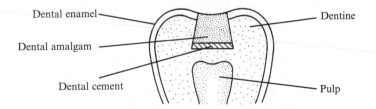

Dental enamel — Dentine
Dental amalgam
Dental cement — Pulp

Teeth can be restored ...
... by means of an inlay of gold alloy or porcelain ...
... by filling with dental amalgam ...

(b) The cavity is filled with amalgam which can be moulded to fit the cavity exactly and then allowed to set hard. The filling is not bonded to the tooth; the cavity is shaped to prevent the filling from falling out. A layer of dental cement insulates the bottom of the cavity.

... by filling with
a poly(alkenoate)
cement ...
... or a polymer-ceramic
composite.

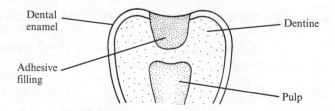

(c) The cavity is filled with a material that bonds to dentine, e.g. a glass poly(alkenoate) cement or a polymer-ceramic composite.

The **dental restorative materials** which are most popularly used today are of three kinds: dental amalgams, glass poly(alkenoate) cements and polymer–ceramic composites. (An alkenoic acid is an unsaturated carboxylic acid, e.g. propenoic acid, $CH_2{=}CHCO_2H$.)

DENTAL AMALGAMS

In the fifteenth century, an Italian, Giovanni of Arcoli, developed the method of cleaning a dental cavity of decayed matter and then filling it with gold leaf. Gold fillings and crowns are still used – and still costly. For most of the population, extraction was the only affordable treatment until 1826, when the first dental amalgam was made by alloying bismuth, lead and tin with mercury. (An amalgam is an alloy which contains mercury.) Mercury is a liquid at room temperature, and amalgamation with other metals makes it solidify at room temperature. The original amalgam had a melting temperature close to 100 °C and had to be poured into the cavity at this temperature. As you can imagine, work continued in the search for an amalgam with a lower melting temperature! The science of dentistry took a big step forward in 1890 when G.V. Black developed an amalgam containing silver, tin, copper and zinc which was liquid at room temperature when freshly mixed but solidified and hardened quickly. The amalgams which are used today are of a very similar composition [see Figure 6.7B].

FIGURE 6.7B
The Composition of the
Standard Dental Alloy

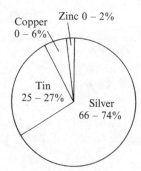

Silver provides resistance to tarnish and strength and slows the hardening of the amalgam. Tin is needed because it has a high affinity for mercury and decreases the setting time. Copper is a strengthening component. Black's amalgam had a high copper content; since the 1960s low-copper amalgams have been used. Zinc is added to reduce the tendency of other metals to oxidise.

Dental alloys are supplied as powder or tablets to be added to an equal mass of mercury and mixed in an amalgamator which vibrates vigorously to give thorough mixing. The amalgam is placed in the tooth, 'carved' (moulded to the cavity), compacted and allowed to harden for 5–10 minutes. The amalgam continues to harden for some months, but 96% of the hardening has occurred within 2 days. On setting, the amalgam expands a little, preventing saliva entering between the filling and the tooth. All dental amalgams are brittle, and fracture is apt to occur at the margins.

Dental amalgams contain mercury alloyed with silver, tin, copper and zinc. They are strong and hard-wearing, but they do not resemble natural teeth.

Dental amalgams form hard, long-lasting fillings. They never look like natural teeth and are used for the restoration of back teeth. Some people are concerned over the release of mercury, but the release of mercury from corrosion of amalgams is 4–20 μg day^{-1}, which is about the same amount as we ingest in our food. Copper salts are toxic, but copper in the amalgam is combined with tin and silver and undergoes very little corrosion.

DENTAL CEMENTS

A dental cement is used to attach an inlay to the surface of a cavity [see Figure 6.8A(a)]. All dental cements are made by reacting a basic powder with an acidic liquid. The powder is in excess. When the powder and the liquid are mixed, an acid–base reaction occurs and a solid cement forms. The set material consists of unreacted powder surrounded by a matrix of reaction product. The most popular cements today are glass poly(alkenoate) cements (see below).

GLASS POLY(ALKENOATE) CEMENTS

Filling materials that resemble natural teeth are the glass poly(alkenoate) cements. They are made by combining a glass powder with a polymer. The glass is an aluminosilicate glass with a high fluoride content. The liquid component is a 50% aqueous solution of a poly(alkenoic acid), e.g. the polymer of propenoic acid, $CH_2{=}CHCO_2H$. On mixing, the acidic polymer reacts with the surface of the glass particles to displace metal ions. These ions dissolve in the liquid and make it set by forming cross-links between carboxylate groups in the polymer chains. The final solid is a composite of glass particles coated with a silica gel in a matrix of metal poly(propenoates). Glass particles may form 70% by volume of the material.

Glass poly(alkenoate) cements use glass particles in a matrix of a poly(alkenoic acid). Chemical reactions take place to bond the glass particles to the matrix. The cements resemble natural teetch.

Glass + Poly(alkenoic acid) $\longrightarrow$ Poly(alkenoate) matrix + Silica gel coating on particles

Glass poly(alkenoate) cements are translucent and resemble natural teeth. They adhere well to dental cavities because their carboxyl groups chelate to calcium ions in dentine. They are used to line the base of dental cavities beneath other restorative materials and as a cement for bonding crowns and inlays [see Figure 6.7A]. However, the extensive cross-linking makes them more brittle than polymer–ceramic composites (see below). When translucency is not required for the sake of appearance, polymer–ceramic cements which are opaque but stronger and faster-setting are used.

POLYMER-CERAMIC COMPOSITES

One way of filling a cavity with a pliable material which hardens inside the cavity is to use a monomer which can be polymerised inside the cavity. The first material to be tried was poly(methyl 2-methylpropenoate). It was not satisfactory because it contracts on setting to leave a gap between the filling and the tooth. It has a coefficient of thermal expansion which is about ten times that of enamel and dentine, and therefore expands more than the tooth in contact with hot and cold food and drink. Poly(methyl 2-methylpropenoate) was superseded by dental composite materials.

Polymer–ceramic composites are composed of ceramic particles dispersed in a matrix of poly(alkenoate) resin. The ceramic particles are often of silica or alumina or a glass. The composite is made by dispersing the ceramic particles in fluid monomer, which includes the initiators required for polymerisation ('curing'). The monomer polymerises ('cures') to a hard solid, while the ceramic particles do not undergo chemical change. The ceramic particles are coated with a silane bonding agent which contains CH_3O— groups which bond to ceramics and $>C=C<$ groups which bond to the polymer. The matrix is chosen for low volatility, low shrinkage on polymerisation, rapid hardening and the production of a strong material. A popular choice is the polymer of $H_3CO_2CCH=CH—CH=CHCO_2CH_3$.

Some polymer–ceramic composites are formulated by manufacturers as a pair of pastes which polymerise on mixing. Others are supplied as a single paste which polymerises on application of intense visible light. The latter gives the dentist more time to mould the paste to the cavity. However, he or she may have to install a large filling in several layers because light will not penetrate beyond a certain depth. About 50 composite resin formulations are available.

Polymer–ceramic composites use ceramic particles in a polymer resin, e.g. a polyester. They resemble natural teeth.

Polymer–ceramic composites do not bond chemically to enamel or dentine, but they can be made to form a mechanical bond. A thin layer of enamel is dissolved with a concentrated solution of phosphoric acid. The enamel is then very rough, and fluid monomer can flow into pits in its surface. As the monomer polymerises, a mechanical bond forms between the polymer and the enamel [see Figure 6.8A (c)]. Sometimes front teeth become damaged or discoloured. The dentist can use this technique to attach a thin layer of ceramic material to the front surface of a tooth.

REPLACEMENT OF TEETH

If a material with a porous surface is placed in contact with bone, then, under the right circumstances, bone may grow into the surface to produce bonding through mechanical interlocking. This process can be achieved with metals, ceramics and polymers. Some materials promote new bone formation at the interface, e.g. calcium phosphate ceramics, including **bioglass** (based on $Na_2O \cdot CaO \cdot P_2O_5 \cdot SiO_2$).

The replacement of an extracted tooth by a tooth implant is more generally available since the introduction of ceramics such as Bioglass which bonds to bone.

An implant can be inserted into the socket left by an extracted tooth. After the implant has bonded to the bone, a crown is built on to the implant. Dozens of different ceramics have been investigated as potential implant materials. Porous ceramics are used so that regenerating bone can grow into the implant. The crown may be formed from a polymer–ceramic composite or a glass poly(alkenoate) as described above. The number of ceramic tooth implants in the USA in 1986 was 300 000; a figure of 1 500 000 is projected for 1996.

6.7.3 TENNIS RACQUETS

A tennis racquet needs a frame and a handle which are strong, stiff, light and tough, can withstand impact loading and do not creep or warp on exposure to changes in temperature or humidity. The materials must be able to be moulded to the required shape. When the ball hits the strings, there is a large impact force which sets up vibrations. Unless the player hits the ball in the centre of the racquet head with a firm wrist, the vibrations are transmitted through the frame to the player's elbow. The vibrations must be reduced in amplitude or they will damage the elbow and give rise to a painful 'tennis elbow'. This injury affects a large fraction of tennis players. Cost is a factor, although professionals are willing to pay a high price for a good racquet. The materials which combine a high ratio of tensile strength/density (specific tensile strength) and a high ratio of elastic modulus/density (specific elastic modulus) are wood, metal and composites. Some possible materials are listed in Table 6.8B.

The requirements of a tennis racquet are that it should be strong, stiff and tough . . .
. . . and dampen the vibrations which are transmitted from the strings to the player's elbow . . .
. . . and be affordable

Material	Specific strength/ (MPa kg^{-1} dm^3)	Specific stiffness/ (GPa kg^{-1} dm^3)	Relative toughness	Relative vibration damping	Relative cost
Wood (hickory)	105	21	Good	Good	Low
Aluminium–magnesium alloy	54	25	Good	Poor	Medium
Nickel–chromium–molybdenum steel	115	27	Good	Poor	Medium
Composite: epoxy + 60% carbon	890	90	Medium	Medium	High
Composite: epoxy + 70% glass	750	25	Medium	Medium	High

TABLE 6.7B
Possible Materials for Tennis Racquets

Wood is the best material for dampening vibrations . . .
. . . and metals are bad in this respect.

Wood, the traditional material, is a good choice because it is tough and strong relative to its density and has good damping for vibrations. The cost puts it within the reach of the ordinary tennis player. The tendency to warp can be overcome by using laminated wood. The life of a wood and ox gut racquet is about 5000 impacts.

Aluminium alloy was the first material to follow wood. A length of tubing is shaped into a frame. Aluminium alloy has the advantage over wood of being stiffer, but it has poorer damping properties and is more expensive. After about 6000 impacts an aluminium alloy frame shows some plastic deformation.

Steels offer higher strength and stiffness than wood or aluminium. They have poor vibration damping and they can become corroded.

Composite materials, e.g. fibre-reinforced epoxy resins, are low in density and very stiff and give the player an advantage.
A special method of construction must be employed to fill the frame with material that dampens vibrations . . .
. . . making the racquets expensive . . .
. . . but they are the choice of professionals.

The combination of properties needed in a tennis racquet (and in squash and badminton racquets also) can best be met by composite materials. Carbon-fibre-reinforced epoxy resin and glass-fibre-reinforced epoxy resin have very high strength/density ratios and high stiffness, especially with carbon fibres. The high stiffness enables them to impart a large force to the ball. The vibration damping is less good than wood but better than metals. The vibration damping is improved by moulding the frame and handle as a hollow tube and filling it with a polyurethane foam. The composite racquets give the best performance and are chosen by professionals. The new racquets are stronger and lighter and last longer than wood or metal racquets. They are also more expensive. Kevlar fibres, E-glass fibres and ceramic fibres are now being used as fibre reinforcements. A carbon-fibre-reinforced polymer racquet with nylon strings loses only 3–4% of its stiffness after 50 000 impacts. Wood is the next best racquet material, followed by aluminium alloy. Steel is not used.

In the days when ox-gut was used to string tennis racquets, two oxen were needed to provide enough string for one tennis racquet. If it were not for the development of

Synthetic fibres, e.g. nylon and Zyex, are used for stringing racquets.

modern synthetic fibres, there would not be enough string to supply the millions of racquets which are sold today. Nylon strings are widely used. Nylon has a higher tensile modulus than ox-gut and the rebound velocity of the ball is therefore higher. Each nylon string is a bundle of entwined nylon fibres. Zyex [§ 4.17.5] fibres are also used; they are more resistant to abrasion and water than natural gut and retain tension over a longer period.

FIGURE 6.8C
A Composite Tennis Racquet Frame – one method of construction

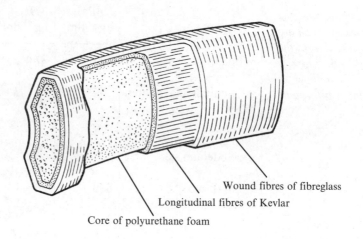

Wound fibres of fibreglass
Longitudinal fibres of Kevlar
Core of polyurethane foam

Replacement organs must be able to withstand physical strains and chemical attack. They must not be rejected by the body . . . . . . they must be biocompatible.

6.7.4   HIP REPLACEMENTS

One solution to the problem is to coat plastic replacement parts with a material that has been derived from the body, e.g. gelatin.

Hip joints wear out in many people, causing pain and lameness. The replacement of hip joints by artificial joints is carried out in tens of thousands of patients each year. The hip joint works like a ball and socket. In a total hip replacement, surgeons cut out the patient's own joint. They insert an artificial socket into the pelvis and the ball has a shaft which is pushed down into the femur, the thigh bone. They cement the implants to the bone [see Figure 6.7D]. The NHS performs 40 000 hip replacements and 16 000 knee replacements a year in the UK.

FIGURE 6.7D
A Cemented Total Hip Replacement

Hip joints wear out in many people, and the replacement of hip joints by artificial joints is performed on thousands of people each year.

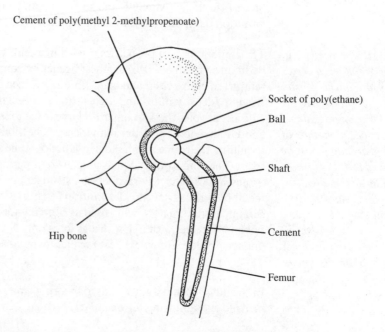

Cement of poly(methyl 2-methylpropenoate)

Socket of poly(ethene)
Ball
Shaft
Cement
Femur
Hip bone

Carbon fibres again

Carbon fibres are used to repair damaged tendons and ligaments. The fibres are coated with a biodegradable polymer to help to prevent fraying or snapping. They are used to connect tissues and also, with the help of a biodegradable rivet, bones. They are used to repair injuries to knee joints and ankles. As new tissue grows round the carbon fibre, the artificial material is absorbed into the body.

Replacement bones

Composites of carbon fibres and polymers look promising as materials for artificial bones. When metal implants are used to replace bones, they take the load off local bone. Normal weight-bearing exercise encourages bone formation. When there is no weight bearing, bone tends to weaken and deteriorate. Composites are better than metals as artificial bones in that they are flexible and place the load on local bone, which thrives and regenerates under stress.

What are the requirements for an artificial hip?

An artificial hip joint must be ...
... corrosion-resistant, non-toxic, biocompatible, strong, tough and fatigue-resistant.

1. It must be corrosion-resistant. It is essential that the material is not attacked by body fluids, not only to ensure a long life for the prosthesis but also so that no harmful substances enter the bloodstream.

2. It must be non-toxic. No harmful substances must be released from the prosthesis into the body.

3. It must be biocompatible. When a foreign substance is implanted in the body, sometimes the body rejects the substance. Other substances are tolerated by the body: they are biocompatible.

4. It needs strength, toughness and fatigue-resistance. If bits wear away, the debris which accumulates causes friction in the joint and this may accelerate wear. When an artificial hip fails on one of these grounds, it has to be replaced, with discomfort and pain to the patient.

The hip joint is a ball and socket joint. The ball and the shaft which is inserted into the femur are made of an alloy ...

Of the materials available, a metal or an alloy is chosen for the ball and shaft. Polymers are easily shaped and many are biocompatible, but they are not strong enough. Ceramics are hard-wearing and strong, low in density and do not corrode. Many are biocompatible, and new techniques allow many of them to be moulded into the required shapes. An advantage of a ceramic implant is that it can be made porous so that bone can grow into it. The brittleness of ceramics precludes their use at present, but there is a possibility that research will result in the invention of new ceramics which can be used.

... e.g. stainless steel, cobalt-chromium alloy, titanium alloy ...

The alloy most often used is stainless steel. This lasts well in elderly people but when a younger, more active person gets a hip joint the stresses which he or she puts on it may wear out the joint in five years. Stainless steel is rigid, whereas bone bends a little. As a result bone may bend around the implant and loosen the bond between bone and implant. A cobalt–chromium alloy is another material in standard use, more resistant to corrosion and easier to cast than stainless steel. Titanium is too soft to use alone, but an alloy of titanium with aluminium and vanadium is used. It has some degree of flexibility which allows it to move with bone.

For the socket, poly(ethene) is chosen. It provides a smooth surface for the ball to move against. Constant use wears away the surface and particles of poly(ethene) accumulate in any gaps between the implant and the bone, causing inflammation which makes the implant work loose.

Poly(methyl 2-methylpropenoate) is chosen for the cement. It is prepared under reduced pressure to avoid bubbles, fired into the joint with a gun and moulded to maintain contact with both bone and implant.

A new technique which is the subject of research is to dispense with cement and allow new bone to grow into the metal implant. Titanium is biocompatible, and bone grows into it – a process called **osseointegration** – to form a bond which is stronger than cement. The biocompatibility of titanium is due to the unreactive layer of oxide that forms on its surface. It does not provoke a reaction from the immune system and allows normal bone repair to take place so that there is a secure bond between implant and bone. The implant is coated with a surface that encourages osseointegration. The technique has been used with titanium implants in the head for people who have lost ears or nose through accident or disease and for artificial crowns and dentures. It has not been used in hips and knee joints because titanium is too soft for such weight-bearing joints.

6.7.5   BICYCLES

FIGURE 6.7F
An aluminium alloy is chosen for this light mountain bike

The bicycle is the most efficient machine for converting energy into motion. The first bicycle frames were made of wood. Bamboo frames were still in use until the end of the nineteenth century, when they were superseded by tubular steel frames. The chief materials used today are (i) low-carbon steel and medium-carbon steel for roadsters, and (ii) chromium–molybdenum–manganese steels for competition cycles. A steel-frame bicycle with a mass of 2–3 kg can carry 100 times its own weight. This safety factor is rarely achieved in other forms of construction, e.g. bridges, aircraft and cranes.

Bicycle frames are for the most part made from steel.

Other materials in use are aluminium alloys, magnesium alloys, titanium alloys and carbon-fibre-reinforced epoxy resin.

Other materials include . . .
. . . magnesium alloys, which can be cast in one piece . . .
. . . titanium alloys for very lightweight machines . . .

● The whole frame can be cast in one piece from a magnesium alloy. This frame is very light because magnesium has only 1/3 the density of aluminium and because the need for nuts and bolts to join the sections together is avoided.

● Titanium is more expensive than steel but it is used in top-of-the-range mountain bikes because of its low density, high fatigue strength, high toughness and high corrosion resistance. The density of titanium is $4.5\,kg\,dm^{-3}$ and that of steel about $7.8\,kg\,dm^{-3}$. Titanium alloy bikes are light, strong and fast.

. . . all-polymer cycles which can be made by injection moulding . . .
. . . fibre-reinforced composites . . .
. . . with fibres of carbon or Kevlar or ceramic . . .
. . . and a matrix of polyester resin or epoxy resin . . .

● All-polymer cycles for everyday use have been made with wheels of glass-fibre-reinforced polyamide and a frame of glass-fibre-reinforced polyester which can be made in one piece by injection moulding. The all-polymer cycles look rather cumbersome compared with tubular steel cycles [see Figure 5.9A].

● Fibre composites are used, employing fibres of carbon for stiffness or Kevlar for toughness or ceramic for damping resistance. The resin may be a polyester resin or an epoxy resin. Such frames can be cast in one piece, like Chris Boardman's 'monocoque'. Many top-class long-distance riders use carbon-fibre-reinforced resin bikes for mountain stages because of their lightness and titanium bikes for other stages of a race.

● A metal matrix with ceramic fibres has been used. The matrix is a copper–aluminium alloy, and the fibres are silicon carbide ceramic. It is an extremely expensive material, but it has the advantage over polymer frames that it is easier to join the components.

. . . which make the very lightweight bikes that are popular with long-distance riders.

The frame is not the only part of the bicycle to change. The off-road bikes need more rigid wheels than road bikes and glass-fibre-reinforced nylon has been tried out in wheels. In competition cycles disc wheels have been used, in which a disc of aluminium alloy or of carbon-fibre-reinforced composite replaces the spokes in the wheels. The disc wheels are aerodynamically better than wire-spoked wheels. Also in use are three- and five-spoked wheels, which have the advantage over disc wheels in cross-wind conditions [see Figure 5.9A].

Disc wheels are used in some competition cycles.

On cost grounds it is unlikely that steel-frame cycles will be overtaken by the new composite materials. On the competition scene, however, riders are prepared to pay a high price for a lighter bicycle and for a model with less wind resistance. The best carbon-fibre-reinforced frame weighs only 1–1.5 kg, compared with 2–3 kg for a steel frame. A titanium alloy frame is lighter than a steel frame and heavier than a carbon-fibre composite and is stronger than a carbon-fibre composite frame.

QUESTIONS ON CHAPTER 6

1. A fizzy drink container may be made of metal, glass or plastic.

(*a*) List, with reasons, the properties which such a container must possess.

(*b*) State the relative merits of the three materials for use in such a container.

2. (*a*) List the properties that would be required of materials chosen for the following applications: (i) a telephone, (ii) a kitchen work surface, (iii) a bucket, (iv) a car bumper, (v) a car axle, (vi) a step ladder, (vii) a vehicle engine block.

(*b*) Name a material that is used for each of these applications and explain why its properties fit it for the purpose.

3. The figure shows a cross-section of a total hip replacement.

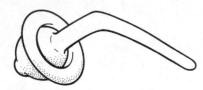

(*a*) State three properties required by this replacement, and give a reason for each property.

(*b*) At present metals and polymers are used for the 'ball' and 'socket' of the hip replacement. Research work has shown that ceramics may serve well in this function. Compare the characteristics of metals, polymers and ceramics for use as implants.

4. The figure shows a load-extension curve for (i) cold-worked copper, (ii) annealed copper, (iii) a ceramic, (iv) a polyamide.

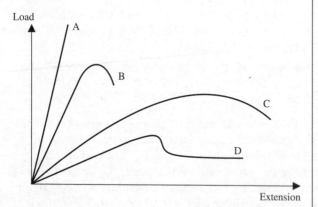

(*a*) State which curve is for which material.

(*b*) Beams made of each of the four materials with the same square cross-section were fixed horizontally at one end and loaded at the other. Describe what will happen to each beam under load.

(*c*) Which of the four beams would be the most serviceable for operating in a climate with extremes of temperature and bad atmospheric pollution?

5. State four important properties required of a material that is used for filling front teeth. In each case explain why the property is important.

6. (*a*) Explain the advantages of carbon-fibre-reinforced nylon for the manufacture of tennis racquet strings.

(*b*) Compare the use of wood, aluminium and composites for tennis racquet frames.

7. How do the following materials achieve a resistance to corrosion?

(*a*) a carbon steel with a coating of paint

(*b*) aluminium

(*c*) copper exposed to air

(*d*) nickel and nickel alloys in air

(*e*) titanium and its alloys in air

(*f*) plastics, e.g. poly(ethene)

(*g*) ceramics, e.g. silica

8. Parts of the human body can be replaced by synthetic materials. Such parts must be strong, flexible, harmless and must not be rejected by the body. Explain what properties make the following materials suitable for their uses.

(*a*) an artificial arm bone of aluminium.

(*b*) an artificial artery of dacron polyester knitted into a narrow tube

(*c*) a heart valve of poly(ethene)

(*d*) an artificial arm bone of carbon-fibre-reinforced plastic

(*e*) fillings in teeth of an amalgam of mercury with other metals

(*f*) an ear bone of teflon

(*g*) an artificial hip with a ball-and-shaft of stainless steel and a socket of poly(ethene)

9. You are asked to choose the best material for a vaulting pole. The pole must not buckle under the load of the vaulter, it must be stiff, must bend elastically but not deform plastically, and it should be as light as possible. Engineers calculate that the criterion to use for comparing the stiffness of materials is the value of $\sqrt{E}/\rho$, where E = tensile modulus and ρ = density. By using the values in the table, say which material is the best (with the highest value of $\sqrt{E}/\rho$) and say how it compares with bamboo which was the original choice for vaulting poles.

Material	Tensile modulus, E/GPa	Density, ρ/(kg dm^{-3})
Aluminium	73	2.8
Bamboo	10	0.5
Carbon-fibre composite	200	2.0
Magnesium	42	1.7
Steel	210	7.8
Titanium	120	4.5

ANSWERS TO SELECTED QUESTIONS

CHAPTER 1: PROPERTIES OF MATERIALS

Checkpoint 1.3
1. 100 MPa
2. 3 MPa
3. 1%
4. 0.03%

Checkpoint 1.5
1. C
2. 15 kN
3. 240 kPa
4. Can bear a compressive load but cannot bend, e.g. a vehicle chassis.

Checkpoint 1.6
1. C
2. 25 GPa
3. 1.0×10^{-3}
4. (a) 120 GPa (b) 420 MPa

Checkpoint 1.7
1. See §1.7.1.
2. (a) See §1.7.1 (b) see §1.5.1
3. (a) See Figure 1.7D (b) For a test piece of dimensions 10 mm × 10 mm, Energy required to break the piece = (Mass of pendulum × g × Difference in heights) = 30 J.
4. (a) All the atoms in the macromolecular structure are covalently bonded together; see §1.7.3
   (b) Quartz has a macromolecular structure; calcium carbonate has an ionic structure
   (c) Calcium sulphate has an ionic structure; talc has a layer structure; see §1.7.3.

Checkpoint 1.11
1. (a) A ductile fracture occurs in the ductile temperature range; see Figure 1.11C, and a brittle fracture occurs below the transition temperature
   (b) Brittle fracture is illustrated in Figure 1.11D(a) and ductile fracture in Figure 1.11D(c).
2. A notch concentrates and multiplies the chances of failure; see §1.10.2.
3. See Figure 1.11C.
4. (a) See §1.9 (b) see Figure 1.9A (c) see Figure 1.9B

Checkpoint 1.12
1. Fatigue failure is failure due to repeated application of a stress which is too small to fracture the specimen when applied once only.
2. Corrosion, overstressing, accidental damage, cracks.
3. Calculation of permissible size of cracks as described in §1.12.
4. C

Checkpoint 1.15
1. 8.0 Ω
2. 4×10^4 V
3. (a) high tensile strength and high melting temperature
   (b) high tensile modulus
   (c) high tensile strength and resistance to corrosion
   (d) high electrical conductivity and high melting temperature
   (e) high impact strength
   (f) high fatigue limit
   (g) high electrical conductivity, high ratio of tensile strength/density, resistance to corrosion.

Questions on Chapter 1
1. (a) tensile strength (b) ductility
   (c) electrical conductivity (d) thermal conductivity
   (e) toughness (f) fatigue resistance (g) hardness.
2. See Figure 1.5B.
   Yield stress = Force needed to cause a permanent deformation (the specimen does not return to its original length when the force is discontinued).
   Tensile strength = Highest stress that the specimen can bear; beyond this it elongates to fracture.
   When the load increases beyond the maximum stress or tensile strength, a specimen becomes too unstable to support the load and fractures at a value of stress called fracture stress. Fracture stress is less than tensile strength because the specimen loses work hardening due to deformation.
3. (a) See Figure 1.5D (b) see Figure 1.5F
   (c) see Figure 1.5E
4. (a) 1270 MPa (b) 35 MPa
5. B
6. See §1.12.
7. The material (a) can withstand the application of a high stretching force without breaking, (b) is stiff,
   (c) needs only a small stress to make it reach the point where it begins to stretch without further application of force (Figure 1.5B), (d) is easily dented.
8. (a) 20 MPa (b) 5×10^{-3}
9. (a) A = yield stress, at which plastic deformation begins
   (b) a line parallel to AO
   (c) C is the highest stress which the material can bear; it is greater than that at A because the material has been work hardened.
   (d) (i) Fatigue failure, see §1.12; possibly because cracks develop (ii) testing as in §1.12.1.

CHAPTER 2: METALS

Checkpoint 2.4

1. (*a*), (*b*) See § 2.3
   (*c*) See § 2.3 and Figure 2.3A. When the arrangement of cations changes to the new shape, the attraction between the cations and the electron cloud is unchanged.
   When a potential difference is applied between the ends of the metal, the delocalised electron cloud flows towards the positive potential.

2. Many transition metals have a large number of electrons in the outer valence shell and therefore a large delocalised electron cloud.

3. See § 2.4, first paragraph.

4. (*a*) steel alloyed with molybdenum, tungsten or titanium
   (*b*) brass   (*c*) nickel–copper alloy, called nickel silver
   (*d*) nickel–chromium steel
   (*e*) aluminium alloys, e.g. Duralumin, titanium alloys
   (*f*) tin–lead solder.

Checkpoint 2.7

1. (*a*) A single crystal is formed by crystallisation about a single nucleus, and all the atoms or ions occupy perfectly regular positions in the same 3-D structure.
   (*b*) A polycrystalline solid consists of a mass of individual crystals. The axes of individual crystals are different from those of neighbouring crystals; see § 2.6.

2. (*a*) A dislocation is a defect in a crystalline structure, see § 2.7.2 and Figure 2.7B
   (*b*) The movement of a dislocation across a structure involves less bond-breaking than the movement of one plane of atoms relative to another. The ductility and malleability of metals are therefore much higher than calculated on the block slip theory.

3. Cold working involves distorting the structure and forcing dislocations to move. Cold working increases the number of dislocations. The more dislocations are present, the more they get in the way of each other and impede movement. When the dislocations can move only with difficulty, the metal has lost ductility and malleability and become harder.

4. e.g. cold working, reducing the temperature, adding grain-refining agents to produce a fine-grained metal which is harder than a coarse-grained metal.

Checkpoint 2.9

1. (*a*) crystals of eutectic of composition 0.45 Cd; 0.55 Bi and crystals of Bi
   (*c*) eutectic mixture of Si + Al, containing 1.00 mol Al + 0.124 mol Si, and 0.876 mol crystals of Si
   (*d*) (i) 15.1 g of eutectic mixture   (ii) 1.0 g Si crystallises; then 15.1 g of eutectic solidifies
   (iii) 2.0 g Be crystallises; then 10.1 g of eutectic.

2. (*a*) (i) Tin crystallises   (ii) Eutectic solidifies
   (*b*) See § 2.9.2
   (*c*) Large crystals of tin and fine crystals of tin and lead.

Checkpoint 2.10

1. See §§ 2.10.2, 2.10.3, 2.10.4, 2.10.6.

2. See § 2.10.6.
   (*a*) Some recrystallisation of austenite as ferrite, making the steel less brittle
   (*b*) Hard and malleable.

3. (*a*) The bcc structure is less close-packed [see § 2.5]. On rapid cooling, carbon is trapped in the austenite structure, distorting the structure, impeding movement of dislocations and making the metal hard and brittle.

4. (*a*) Less mechanical energy is required. Annealing takes place.
   (*b*) Less heat energy is required. Small accurate changes in shape can be accomplished.

Checkpoint 2.11

1. Sand castings cool slowly and have larger columnar crystals and relatively low strength. Die castings cool more rapidly and have equiaxed crystals and higher strength.
   (*a*) Die casting; fine-grained
   (*b*) Pressure die casting is good at forcing metal into all parts of a complicated mould
   (*c*) In sand casting, a riser holds liquid metal [Figure 2.11A]. In die casting, gravity or pressure or centrifugal force is utilised to make the metal fill the mould; see § 2.11.1
   (*d*) Investment casting, using a ceramic mould, see § 2.11.1
   (*e*) Sand casting.

2. Rolling [Figure 2.11E], drawing [Figure 2.11F], pressing [Figure 2.11H], deep drawing [Figure 2.11G], sand casting [Figure 2.11A] or forging [Figure 2.11I].

Checkpoint 2.13

1. Water is needed for rusting; see equations in boxes in § 2.12.

2. Oxygen is needed for rusting; see equations in boxes in § 2.12.

3. The metals will corrode more the further apart they are in the electrochemical series.

4. Aluminium and copper are more widely separated in the electrochemical series than aluminium and zinc or aluminium and iron; galvanised steel is the better choice.

5. Magnesium corrodes in preference to iron; see sacrificial anodes in § 2.13.

6. Iron is higher in the electrochemical series than copper, so iron corrodes in preference to copper.
   Aluminium is above iron in the electrochemical series so it becomes a sacrificial anode, corroding in preference to iron.

7. e.g. galvanising, attaching a sacrificial anode.

8. Diffusion of oxygen from air to the metal through structural defects in the oxide film; see § 2.11.

Questions on Chapter 2

1. Delocalised electron cloud; see § 2.3.

2. (*a*) For metallic bond see § 2.3.
   (*b*) Dislocations (§ 2.3) enable deformation to take place more readily than by metal block slip.
   (*c*) For grains see § 2.6. The movements of dislocations are hindered by grain boundaries [§ 2.7.2]; therefore a fine-grained metal is harder and less ductile than a coarse-grained metal.
   (*d*) Imperfections in the crystal structure hinder the movement of dislocations and reduce ductility; see Figure 2.7E.

3. See § 2.10.
4. (a) Use e.g. Brinell hardness test [§ 1.7.1].
   (b) Check temper colour; pale straw for untempered to dark blue for tempered. Test tensile strength; tempered steel is more ductile. Test hardness; quenched but untempered steel is harder [§ 2.10.6].
   (c) Test the tensile strength; § 1.5.1.
5. See §§ 2.4, 2.8, 2.9.
6. Yield stress for metals which do not show much work hardening; see Figure 1.5E.
   Fracture stress for brittle materials which fracture soon after the elastic limit; see Figure 1.5D.
   Tensile strength for e.g. metals which can be work-hardened so that tensile strength is greater than yield stress and for polymers which can be cold-drawn [Figure 1.6A].
7. (a) The liquid cools until it crystallises suddenly and completely at the eutectic temperature to form crystals of the eutectic mixture.
   (b) Crystals of B form. When the eutectic temperature is reached, the remainder of the liquid solidifies as the eutectic mixture.
   Appearance: (a) fine crystals of A and B,
   (b) coarse crystals of B and fine crystals of A and B.
8. (a) See § 2.12.1
   (b) See § 2.13

(c) (i) a coating of a sacrificial metal
   (ii) application of an external voltage.
9. (a) (i) fcc (ii) bcc.
   (b) See Figures 2.5A and B
   (c) See § 2.10.6.
10. (a) See § 2.8
    (b) See Figure 1.5B
    (c) See Figure 2.7A.
11. (a) High temperature and corrosive atmosphere
    (b) High tensile strength at high temperature, resistance to corrosion, able to be cast, not too costly
    (c) Carbon steel, stainless steel, copper and aluminium could be used. Aluminium and copper would resist corrosion but are both expensive. Stainless steel is strong and corrosion-resistant, and less expensive than copper or aluminium. Carbon steel becomes corroded after two years or so.
    (d) Carbon steel is generally used, but expensive cars have stainless steel exhausts.
12. (a) Good: aluminium never corrodes
    (b) Impossible: the tungsten is dispersed in action
    (c) Good: easy to collect
    (d) Impossible: zinc is sacrificially corroded
    (e) Impossible to separate from other components
    (f) Good, but disassembly is required.

CHAPTER 3: CERAMICS

Checkpoint 3.3

1. See § 3.1.1 for definition and §§ 3.1.1, 3.1.2 for examples.
2. (a) Hard, good compressive strength, low tensile strength, brittle, low electrical and thermal conductivity, low toughness, high melting temperature; see § 3.1, 3.1.2. Fibres have special properties; see § 3.1.2
   (b) Strong in compression, used as building bricks, drain tiles, etc. and weak in tension, therefore brittle, e.g. china, pottery, glass, etc.
   (c) Eliminate surface cracks by drawing into fibres. Use as a component of a composite material. Reinforce with fibres; see § 3.1.2.
3. (a) Si, O, H, Al (b) Si, O, Al, K
4. (a) Lower glass-making temperature; see § 3.2
   (b) Increase temperature resistance of the ceramic; see § 3.2
5. A vitreous matrix and crystals

Checkpoint 3.4

1. Whiteware – see § 3.4; building bricks – see § 3.4.2; furnace linings – see § 3.4.2.
2. Heat resistance; see §§ 3.4.4, 3.4.9.
3. See §§ 3.4.5, 3.4.6, 3.4.7.
4. Ceramics never corrode, have low density and can be made porous to allow bone to grow into them.

5. e.g. Furnace for smelting a metal – higher melting temperature than metals.
   Machine tools – harder than most metals and alloys.
   Covering for space shuttle – thermal insulation of ceramic fibres.
   Insulators in electrical circuits – good electrical insulators
   Ceramic parts in diesel engines enable the engine to operate at higher temperature, with increased efficiency.
   Ceramic gas turbine blades can operate at higher temperatures than metals.

Checkpoint 3.5

1. (a)

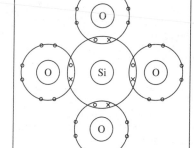

(b)

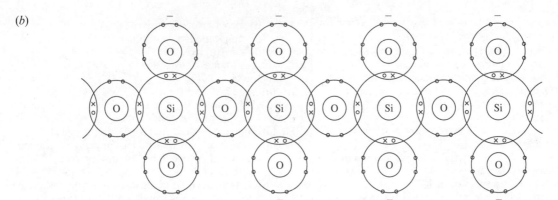

2. (a) Glasses fit the definition, see § 3.1.1
   (b) The temperature at which a ceramic changes from a crystalline structure to the amorphous structures of a glass; see Figure 3.5.I.

3. For structures see Figures 3,5A, B and G
   For silicon(IV) oxide see § 3.5.4. The replacement of some silicon atoms by aluminium atoms, or other atoms, and associated cations gives rise to minerals similar to silica.

4. (a) When molten silica is cooled very rapidly, it solidifies as a disordered glass instead of a crystalline solid.
   (b) See structures in Figures 3.5G and H.

5. (a) Changes shape without fracture; see the metallic bond, § 2.4.
   (b) Fractures: there are few slip planes in an ionic crystal; therefore deformation often brings cations into contact with cations, anions into contact with anions and repulsion occurs.
   (c) Changes shape because the layer structure permits this.
   (d) Fractures: there are no slip planes because of the disordered arrangement of atoms.

6. The amorphous structures of glass means that there are no regular planes of atoms to reflect light; see § 3.5.7.

7. Ceramics are (a) harder and (b) unable to slip because the bonds are directed [see § 3.5.6] and (c) have no delocalised electrons to conduct electricity.

8. (a) For ionic compounds, see Figure 3.5J; for metals see Figure 2.7A
   (b) Ceramics have directed bonds and cannot slip; they are brittle.

Checkpoint 3.6

1. High compressive strength, low cost, convenience of formulation on site, possibility of improving tensile strength by reinforcement [§ 3.6.3] and modifications [§ 3.6.2].

2. See § 3.6.3.

3. (a) Steel or steel-reinforced concrete is necessary for the tensile strength required.
   (b) Concrete is cheaper and a better thermal insulator.
   (c) Concrete is cheaper and a better thermal insulator, and compressive strength is needed, not tensile strength as in (a).

4. (a) (i) More resistant to corrosion and wear, similar to natural teeth
       (ii) Harder and similar to natural teeth
   (b) (i) Harder and more resistant to corrosion
       (ii) Harder and do not creep
   (c) (i) Harder, lower in density, more resistant to corrosion and to creep,
       (ii) Harder, more resistant to corrosion and to creep

(d) (i) High melting temperature, therefore maintain shape at high temperature; low density
    (ii) Much higher melting temperature and vastly superior performance at high temperature.

5. Bronze pots were better thermal conductors and less brittle.

Checkpoint 3.11

1. Annealing of metals is restoring ductility to work-hardened metals by heating to achieve recovery and recrystallisation, see § 2.11.4. Annealing of glass is heating to allow plastic flow to reduce internal strain. It is similar to recovery in metals: no recrystallisation is involved so it differs from annealing in metals; see § 3.8.

2. Float glass process; see Figure 3.7A.

3. (a) High density, high refractive index
   (b) Low coefficient of expansion, high resistance to shock and chemical attack – Pyrex glass
   (c) Coloured glass; see § 3.11
   (d) Photochromic glass; see § 3.11.

4. Laminated glass: sheets of glass glued to a sheet of plastic between them. Tempered glass: made by annealing to reduce stress; see § 3.11.

5. A glass ceramic is a glass which has been devitrified by adding nuclei to induce crystallisation; see § 3.11.2. Properties are closer to ceramics than to glasses, but crystallinity makes them more rigid than glasses.

6. Reinforcement of concrete; see § 3.6.3 of plastics; see § 5.5.

7. (a) See § 3.11.2 and Figure 3.11A
   (b) See Figure 3.11B.

Questions on Chapter 3

1. Many possible examples; see § 3.6.4 for properties.

2. (a) A glass consists of an amorphous arrangement of covalently bonded atoms. A solid electrolyte consists of a regular three-dimensional structure of ions.
   (b) The temperature at which a crystalline ceramic changes to a glass; see Figure 3.5I
   (c) When a liquid ceramic is cooled slowly it crystallises at the melting temperature to form a solid. When a liquid ceramic is cooled rapidly, it is unable to organise the atoms into the regular structure required for crystallisation, and it solidifies as a disordered arrangement – a glass; see Figure 3.5I.

3. Many examples; for properties see § 3.11.

4. (a) Figures 3.5C, D, E and F
   (b) 4
   (c) No metal cations; see Figure 3.5G
   (d) See § 3.10
   (e) Crystallisation or partial crystallisation.

5. (a) Strengthened by removal of surface flaws
   (b) Stress concentration; see §§ 1.10.1, 1.10.2

6. See § 3.12.

CHAPTER 4: POLYMERS

Checkpoint 4.3

1. (a) See §4.3.1
   (b) Thermosoftening plastics consist of individual molecules held together by intermolecular forces of attraction. Thermosetting plastics consist of chains cross-linked by covalent bonds.
   (c) When a high melting temperature is needed, e.g. kitchen work surface, electrical plugs and sockets
   (d) See §4.3.1.
2. (b) See §4.3.2 and substitute —CO_2H in poly(propenoic acid) for —CH_3 in poly(propene).
   (c) A difference in crystallinity; see discussion of poly(propene)

3. CH_2=CH—CH_2OH, $\left(CH_2-CH\right)_n$ with CH_2OH substituent

Checkpoint 4.4

1. (a) See §1.5.2
   (b) When chains reach a certain length, the molecules experience forces of attraction at many points along their length bonding them to neighbouring molecules. The alignment is much greater than with the same number of attractive forces between large numbers of small molecules.
2. Once the substance has assumed the form of aligned molecules, an increase in the length of the molecules does not greatly change the structure of the material.
3. (a) See §1.5.2
   (b) The polymer chains pack together with a higher degree of crystallinity in linear poly(ethene) than in branched-chain poly(ethene) and the higher degree of crystallinity gives the material a higher tensile modulus (makes it stiffer).

Checkpoint 4.5

1. (a) Temperature at which polymer changes from a rigid material to a flexible material; see §4.5
   (b) Flexible, rubbery with low tensile modulus above the glass transition temperature; stiff with high tensile modulus below the glass transition temperature
   (c) Change in molecular motion with temperature; see §4.5.2
   (d) The plastic which is intended to be stiff may become too flexible to perform its function if used above T_g. The plastic which is intended to be flexible may become too stiff for use below T_g.
   (e) Measure tensile strength over a range of temperature; see §1.4.

Checkpoint 4.7

1. (b) Easier to control
2. (a) RCH_2—$\dot{C}H_2$
   (b) This free radical can add to another molecule of ethene:

$$RCH_2—\dot{C}H_2 + CH_2=CH_2 \longrightarrow RCH_2—CH_2—CH_2—\dot{C}H_2$$

   (c) Two free radicals may combine or disproportionate:

$$2RCH_2\dot{C}H_2 \longrightarrow RCH_2CH_2CH_2CH_2R$$
$$2RCH_2\dot{C}H_2 \longrightarrow RCH_2CH_3 + RCH=CH_2$$

   (d) PVC is harder and stiffer than PE. PVC is more polar and the forces of attraction between chains are stronger. PVC has a higher T_g than PE.
4. See §4.7.

Checkpoint 4.11

1. (a) See §4.8.1 for ldpe and §4.9.1 for hdpe
   (b) See Figure 4.9A
   (c) For uses, see §4.8.2, 4.9.2.
2. (a) The greater alignment of molecules in hdpe increases tensile strength.
   (b) It is made by a batch process, compared with a continuous process for ldpe, and incurs the cost of catalyst and solvent.
   (c) See §4.8.1 for structure and §4.9.1.

3. (a) (i) $\left(CH_2-CH\right)_n$ with C_6H_5 substituent   (ii) $\left(CF_2\right)_n$

   (iii) $\left(CH_2-C\right)_n$ with CH_3 and CO_2CH_3 substituents
   (b) For uses, see §4.11.

Checkpoint 4.12

1. (a) See §4.12.1
   (b) Unsaturation; see §4.12.2 or esterification with a triol; see §4.12.1
   (c) Cross-linked polymers are harder, tougher, higher melting, thermosetting.
2. (a) See §4.12.3
   (b) Nylons
3. (a) Phenol has three reactive positions, the 2-, 4- and 6- positions in the ring.
   (b) See §4.12.4
4. (a) and (b) See §4.12.5

Checkpoint 4.15

1. (a) In a batch process, a certain quantity of reactant is allowed to react and then the process is shut down while the reactor is emptied of product and prepared for a fresh batch of reactant. In a continuous process, reactant is fed into the reactor non-stop, and product is led off non-stop, without shutting down the reactor.
   (b) See §4.13
   (c) See §4.13 (1)
   (d) See §4.13 (1). Termination of chain reactions is hindered by increasing viscosity as polymer accumulates in the reactor.
   (e) The solvent reduces viscosity.
2. See §4.14 (7).

3. See § 4.14 (1–6).
4. (*a*) (i) Extrusion moulding, Figure 4.15B
       (ii) Blow moulding, Figure 4.15C and extrusion
       stretch blow moulding, Figure 4.15D.

Checkpoint 4.17
1. See Figure 4.16A.
2. See § 4.17.1.
3. Weaving and knitting; see § 4.16.2.
4. (*a*) Cold-drawing; see § 4.17.1
   (*b*) Alignment of molecules in crystalline regions,
       Figure 4.17A
   (*c*) Increases tensile strength, lustre and resistance to
       absorption of moisture
   (*d*) Increases difficulty of dyeing
   (*e*) Does not absorb perspiration
   (*f*) Texturising; see § 4.17.1
   (*g*) Better for water-repellant fabrics, e.g. waterproof
       jackets.
   (*h*) Amide
   (*i*) Hydrogen bonding; see Figure 4.17B.
5. (*a*) Many hydrogen bonds at frequent intervals along the
       length of the molecule; Figure 4.17C
   (*b*) For uses, see § 4.17.2.
6. (*a*) Terylene; § 4.12.1
   (*b*) Acrilan, Courtelle; § 4.17.4.

Checkpoint 4.18
1. (*a*) A regenerated fibre is made by chemically modifying
       a natural fibre. A synthetic fibre is made by spinning a
       manufactured polymer into the form of fibre.
   (*b*) There is not enough regenerated fibre to meet demand.
2. For (*a*) and (*b*) see § 4.18.1.
3. For (*a*) and (*b*) see § 4.18.2.

Checkpoint 4.19
1. (*a*) See Figure 4.19A
   (*b*) There is a natural tendency for the entropy of a
       system to increase.
   (*c*) Becomes rigid and brittle, non-flexible
   (*d*) Disulphide bridges are formed; see § 4.19 and
       Figure 4.19B.

2. (*a*) $+CH_2\!-\!\underset{\underset{Cl}{|}}{C}\!=\!CH\!-\!CH_2\!+$

   (*b*) $+CH_2\!-\!\underset{\underset{CN}{|}}{CH}\!-\!CH_2\!-\!CH\!=\!CH\!-\!CH_2\!+$

3. Natural rubber contains double bonds which make it
   reactive towards e.g. oxygen. Poly(ethene) has only single
   bonds.
4. See § 4.19.2 on how the arrangement of CH_3 groups
   affects flexibility of chains.

Checkpoint 4.20
1. (*a*) See § 4.20 (1–5)
   (*b*) See § 4.20 for silicone fluids and silicone rubbers.
2. (*a*) See § 4.20
   (*b*) Chains of —Si—O—Si—O—; compare and contrast
       silicones in § 4.20 and silicates in § 3.6.

Checkpoint 4.23
1. (*a*) Crude oil is fractionally distilled. Some fractions are
       cracked with the formation of ethene, which is
       polymerised.
   (*b*) Millions of years
   (*c*) NO
   (*d*) Micro-organisms do not attack them. Plastic waste
       rots very slowly.

2. (*a*) Good thermal insulator
   (*b*) Possibly only minutes
   (*c*) Becomes non-biodegradable plastic waste
   (*d*) Diffuses into the atmosphere. Eventually contributes
       to the greenhouse effect; this effect is less serious if the
       gas is a hydrocarbon than if it is a CFC.
3. (*a*) (i) Avoids the discomfort of removal and the
       necessity for a visit to a doctor for removal.
       (ii) The bag dissolves in the laundry water so it does
       not spread infection to another batch of laundry.
   (*b*) Many examples, e.g. kitchenware, building materials,
       etc.
4. (*a*) Items which are in use for a short time only, e.g.
       carrier bags, food cartons
   (*b*) Many examples of items which are in use for a long
       time, e.g. clothing, garden chairs.
5. See § 4.23, especially 4.23.5.

Questions on Chapter 4
1. (*a*) $HOCH_2CH_2CH_2CO_2H$
   (*b*) $HO(CH_2)_3CO_2(CH_2)_3CO_2H$ (*c*) $+(CH_2)_3CO_2+$
2. (*a*) (i) $HO_2C(CH_2)_4CO_2H$,   (ii) $ClOC(CH_2)_4COCl$,
       (iii) $H_2N(CH_2)_6NH_2$
   (*b*) $+CO(CH_2)_4CONH(CH_2)_6NH+$
   (*c*) See §§ 4.16.1, 4.17.1
   (*d*) Close-packing of nylon chains. The polar groups
       $>C\!=\!O$ and $>NH$ are involved in hydrogen
       bonding between neighbouring chains so few are free
       to bond to water molecules; see § 4.17.1.
3. (*a*) See § 4.18.1
   (*b*) In the formula of cellulose [§ 4.18], each —OH is
       replaced by —$OCOCH_3$
   (*c*) Rayon is cheaper; also see § 4.18.1.
4. (*a*) See § 4.3.2
   (*b*) See Figure 4.9A
   (*c*) See §§ 4.8.1, 4.9.1.
5. (*a*) Unbranched molecules can be aligned into
       crystalline regions, e.g. hdpe, § 4.9.1. Branched
       molecules with short-chain and long-chain branching
       make the degree of crystallinity in a polymer small,
       e.g. ldpe, § 4.8.1
   (*b*) For properties of hdpe, see §§ 4.9.1, 4.9.2; for ldpe see
       §§ 4.8.1, 4.8.2.
6. (*a*) The plastic is flexible above T_g, rigid and brittle
       below T_g
   (*b*) The molecules are aligned by cold-drawing, with an
       increase in tensile strength.
   (*c*) An increase in the degree of crystallinity raises the
       melting temperature; see § 4.17.3
   (*d*) The reduction in cross-sectional area means that the
       molecules have been highly oriented, and this has
       increased the tensile strength of the material; see
       Figure 1.11D.
7. (*a*) Petroleum oil is fractionally distilled, and fractions
       are cracked to yield ethene.
   (*b*) Similar to PVC; see § 4.6 and substitute $CH_2\!=\!CH_2$
       for $CH_2\!=\!CHCl$.
8. (*a*) (i) Covalent bonds link polymer chains.
       (ii) Two or more monomers have been polymerised
       together.
   (*b*) (i) Measure volume (e.g. by displacement) and mass
       (ii) See § 1.5.1
       (iii) See § 1.6.1
       (iv) See § 1.7.2
       (v) See § 1.7.1

9. (*a*) See § 4.3.1
   (*b*) See § 4.16.1
   (*c*) See §§ 4.8.1, 4.9.1, 4.10.1, 4.13
10. For (*a*) and (*b*) see § 4.19.2
   (*c*) Copolymerisation gives cross-linked synthetic rubbers. Carbon black is added to increase strength; see § 4.19.3.
11. When you stretch slowly the polymer chains have time to unfold and line up in the direction of the stretching force, so that the plastic is cold-drawn. When you stretch suddenly, there is not time for this increase in tensile strength to happen.
12. **A**6, **B**3, **C**2, **D**1, **E**4, **F**5
13. (*a*) Is an electrical insulator, can be cast in the required shape and machined to take terminals, screws etc, has a high melting temperature, is hard and tough.
   (*b*) Test conductance, find melting temperature, test hardness and impact strength, find out cost, consider whether the material can be moulded easily.
   (*c*) Thermosetting plastics, rubbers.

14. **A** (*a*) yes (*b*) yes (*c*) polyamide
   (*d*) This is nylon 66.
   B (*a*) yes (*b*) yes (*c*) polyester
   (*d*) This is Terylene.
   C (*a*) no (*b*) no (*c*) polyalkene
   (*d*) This is poly(propene).
   D (*a*) no (*b*) no (*c*) polyalkene
   (*d*) This is a rubber, isoprene.
   E (*a*) no (*b*) no (*c*) polyarene
   (*d*) This is poly(phenylethene), poly(styrene).
   F (*a*) no (*b*) no (*c*) polyalkene
   (*d*) This is poly(chloroethene), PVC.
   G (*a*) no (*b*) no (*c*) a silicone.
   H (*a*) yes (*b*) yes (*c*) a polyamide
   (*d*) This is a peptide or a protein.
   I (*a*) yes (*b*) yes (*c*) an unsaturated polyester
   (*d*) It will form a polyester resin.
   J (*a*) yes (*b*) yes (*c*) a polyamide
   (*d*) This is 'Kevlar'.
   K (*a*) yes (*b*) no (*c*) a poly-ether-ether-ketone
   (*d*) This is 'PEEK'.

CHAPTER 5: COMPOSITE MATERIALS

Checkpoint 5.4

1. (*a*) When a fibre is drawn, the process orients the molecules and the alignment of molecules in a fibre gives it tensile strength. Drawing also removes defects, e.g. surface cracks and scratches, which can concentrate stress; see § 1.10.1
   (*b*) Plastics do not have high tensile strength. The incorporation of tensile fibres enables them to withstand higher tensile stress.
   (*c*) Continuous fibres have the greatest effect on tensile strength; see § 5.3.1.
2. Matrices: e.g. epoxy resins, polyester resins, metals, concrete, etc. Fibres: e.g. glass, carbon, steel, Kevlar, etc.
3. See § 5.3.4.

Checkpoint 5.7

1. (*a*) Concrete is strong in compression, and the steel bar contributes tensile strength,; see § 5.5
   (*b*) See § 5.5.1.
2. The polyester is flexible but weak. The glass fibres are strong but relatively brittle. The composite is both strong and tough; see § 5.7. Uses: boat hulls, car bodies, panels in aircraft, stackable chairs, etc., because low-density, free from corrosion, easy to mould.
3. (*a*) phenylethene
   (*b*) poly(butadiene)
   (*c*) It is a rubber, and rubber particles block the transmission of cracks.
4. (*a*) propenenitrile
   (*b*) $CH_2=CH-CH=CH_2$
   (*c*) Butadiene lowers the tensile strength. It is a rubber and allows the material to elongate further before breaking. It also makes the material tougher.
5. (*a*) Have a rigid structure [see § 4.12.2] and are therefore brittle.
   (*b*) See § 1.10.1
   (*c*) e.g. boat hulls, cycle frames, tennis racquets, etc.

Checkpoint 5.11

1. (*a*) The wood is strongest in the direction of the grain, so sheets of wood are glued together with the grain directions at right angles in successive layers [Figure 5.9A] to give the material strength in two directions.

   (*b*) e.g. plywood, laminated glass [§ 3.11], metals [§ 5.9].
2. Density, fatigue failure, creep resistance, impact strength, cost; see § 5.11.
3. (*a*) The fatigue failure is determined by the metal matrix, and fibres do not improve it.
   (*b*) Creep resistance is increased; see § 5.11.
4. In a brittle matrix, fibres can increase impact strength. In a hard material of high tensile strength, fibres can stiffen the material, make it less ductile and reduce impact strength; see § 5.11.

Questions on Chapter 5

1. See § 5.2.
2. (*a*) See § 5.12.4
   (*b*) See §§ 5.2, 5.6.
3. (*a*) The orientation of molecules produced by drawing the fibre makes it strong and also removes defects in the structure, e.g. scratches and cracks.
   (*b*) (i) easier to mould, harder, cheaper
   (ii) lower in density, easier to mould, cheaper, not corroded, easy to maintain
   (*c*) Continuous fibres with a high degree of orientation are strong; see § 5.3.1
   (*d*) It has less impact strength and is less protective in a collision.
4. 1**D**, 2**H**, 3**A**, 4**G**, 5**C**, 6**I**, 7**E**, 8**B**, 9**F**.
5. See § 5.12.1.
6. Carbon particles fit into spaces between the polymer chains so the chains can change position less readily and the rubber is stiffer and tougher.
7. For corrugated cardboard; see § 5.9.1, Figure 5.9B, and for plywood see Figure 5.9A.
8. (*a*) See § 5.2
   (*b*) See § 5.5 and Figure 5.5A.
9. Corrosion from salt spread on icy roads and from acid rain. Reduce porosity of concrete.
10. See § 5.12.2
   (*a*) Composite disc wheels and composite frames
   (*b*) Disc wheels of composites have less aerodynamic resistance than metal spokes. Frames of composites are lighter and can be moulded in one piece.

CHAPTER 6: REVIEW OF MATERIALS

Checkpoint 6.5

1. PVC consists of flexible polymer chains [Figure 4.3B]. A metal has a crystalline structure, which can be deformed without fracture [§§ 2.7.1, 2.7.2]. Glass is brittle [§§ 3.5.5, 3.5.6], and Perspex is a cross-linked thermoset which is less flexible than the thermosoftening PVC [§§ 4.3.1, 4.4].

2. A ceramic is a covalent silicate framework associated with metal cations. It is impossible to break some of the covalent bonds without shattering the whole structure [§§ 3.1.2, 3.5, 3.5.6]. Metals can be deformed without fracturing by block slip and by the movement of dislocations [§§ 2.6.1, 2.7.2].

3. (*a*) When the fibre is drawn, the orientation of molecules that results increases tensile strength. The process removes defects, e.g. cracks and scratches.
   (*b*) (i) cheaper, easier to mould, easier to maintain
       (ii) cheaper, lower in density, free from corrosion, easier to mould
   (*c*) GRP has less impact strength and gives less protection in a crash.

Questions on Chapter 6

1. See § 6.8.1.

2. (*a*) (i) low-density, tough, hard, easily moulded, electrical insulator
       (ii) thermosetting plastics
   (*b*) (i) high melting temperature, smooth, high impact strength, resistant to corrosion and staining
       (ii) thermosetting plastics, ceramics (and formerly wood)
   (*c*) (i) resistance to corrosion, hardness, high impact strength, low cost, low density
       (ii) metal, e.g. galvanised steel, plastics
   (*d*) (i) very high impact strength, resistance to corrosion
       (ii) reinforced plastic, metal, e.g. steel
   (*e*) (i) very high tensile strength, resistance to corrosion
       (ii) steel
   (*f*) (i) low density, resistance to corrosion, high tensile strength, high impact strength, hardness
       (ii) wood, metal, e.g. aluminium

   (*g*) (i) high compressive strength, ease of moulding
       (ii) cast iron.

3. See § 6.8.5.

4. (*a*) A ceramic, B cold-worked copper, C annealed copper, D polyamide
   (*b*) A extends by a small fraction, then fractures.
       B extends by a small percentage, then necks and soon afterwards fractures. C extends, reaches a point where it extends further without much more increase in load, then necks and fractures. D extends to a point where it extends further by a large percentage without a further increase in load: it has been cold-drawn.

5. See § 6.8.2.

6. See § 6.8.3.

7. (*a*) Paint excludes air and water.
   (*b*) A protective layer of aluminium oxide forms on the surface.
   (*c*) A layer of copper carbonate copper hydroxide forms on the surface.
   (*d*) A layer of nickel oxide forms on the surface and protects against further attack.
   (*e*) A layer of titanium oxide forms on the surface and protects against further attack.
   (*f*) It does not react with air or water.
   (*g*) It does not react with air or water.

8. In addition to being non-toxic and not corroded by body fluids,
   (*a*) low density
   (*b*) high tensile strength, can be obtained in fibre form
   (*c*) flexible
   (*d*) flexible, places load on the bone to help it to regenerate
   (*e*) low melting temperature, expands slightly on setting but not enough to crack the tooth
   (*f*) low-friction, long-lasting
   (*g*) The ball and shaft are strong, tough and resistant to fatigue. The socket is a smooth surface for the ball to work against.

9. Values of $\sqrt{E}/\rho$ show that carbon-fibre composite is the best material (7.1 GPa$^{1/2}$ kg^{-1} dm^3) with bamboo the original material close behind (6.3 GPa$^{1/2}$ kg^{-1} dm^3).

Index

Further reading

Metals in the Service of Man by W. Alexander and A. Street, Tenth Edition (Penguin, 1994)

Tomorrow's Materials by Ken Easterling (The Institute of Metals, 1990)

The New Science of Materials by J.E. Gordon, Second Edition (Penguin, 1976)

Structures by J.E. Gordon (Penguin, 1978)

The Properties of Engineering Materials by R.A. Higgins, Second Edition (Edward Arnold, 1994)

Introduction to Engineering Materials by V.B. John

Shreves Chemical Process Industries by George T. Austin, Fifth Edition (McGraw-Hill, 1984)

The Chemical Industry by Alan Heaton, Second Edition (Blackie, 1994)

Engineering Materials Technology by W. Bolton, Second Edition (Newnes, 1993)

Advanced Materials for Sports Equipment by K.E. Easterling (Chapman and Hall, 1993)

The Essential Chemical Industry, Second Edition (University of York, 1989)

STEAM, Issue 11 (ICI, 1989) and Issue 15 (ICI, 1991)

Chemistry in the Marketplace by Ben Selinger, Fourth Edition (Harcourt–Brace–Jovanovich, 1989)

Materials Science and Technology, Volume 14: Medical and Dental Materials, edited by D.F. Williams